IMAGES OF WAR

THE KALASHNIKOV IN COMBAT

I0822206

IMAGES OF WAR

THE KALASHNIKOV IN COMBAT

RARE PHOTOGRAPHS FROM WARTIME ARCHIVES

Anthony Tucker-Jones

Pen & Sword
MILITARY

First published in Great Britain in 2012 by
PEN & SWORD MILITARY
an imprint of
Pen & Sword Books Ltd,
47 Church Street,
Barnsley,
South Yorkshire
S70 2AS

ISBN 978 1 84884 579 4

A CIP record for this book is available from the British Library.

Frontispiece: Two Northern Alliance fighters in Afghanistan: one has the 'newer' AKM while his comrade has an RPG-7. (*Via Author*)

Typeset in Gill Sans 12pt by Chic Media Ltd

Printed and bound by CPI Group (UK) Ltd, Croydon, CR0 4YY.

Pen & Sword Books Ltd incorporates the Imprints of
Pen & Sword Aviation, Pen & Sword Family History, Pen & Sword Maritime, Pen & Sword Military, Pen & Sword Discovery, Wharncliffe Local History, Wharncliffe True Crime, Wharncliffe Transport, Pen & Sword Select, Pen & Sword Military Classics, Leo Cooper, The Praetorian Press, Remember When, Seaforth Publishing and Frontline Publishing.

For a complete list of Pen & Sword titles please contact
Pen & Sword Books Limited
47 Church Street, Barnsley, South Yorkshire, S70 2AS, England
E-mail: enquiries@pen-and-sword.co.uk
Website: www.pen-and-sword.co.uk

Contents

Photograph Sources

While this book aims to provide the reader with a very brief history of the AK-47, it more importantly offers a visual record of the day-to-day impact of the weapon on the battlefield up to the present day. It is illustrated with a wealth of archive photos ranging from the Vietnam War to the Coalition's invasion of Iraq and beyond.

Illustrating and mapping the combat evolution of the AK-47 and its very numerous variants and copies is no easy task; as it has been involved in practically every single conflict since the Second World War, it would simply be impossible to show them all. What follows is largely a subjective selection drawn from a wide variety of sources, including the author's own extensive military photo collection.

In illustrating this book the author is greatly indebted to Igor Bondarets, Erwin Franzen and Julian Gearing, who regularly covered the wars in Afghanistan and elsewhere during the 1980s and 1990s, as well as Itzhak Bar-Zait. Igor specialised in the Soviet Army, while Erwin and Julian spent considerable time with the Mujahideen (a separate title in this series on the Soviet–Afghan War will further highlight their work). Their kindness and generosity in assisting with this project is greatly appreciated.

The author would also like to thank the numerous combat photographers of the US Department of Defense, US Army, US Navy, US Air Force and ISAF Media for sharing their work in Afghanistan and Iraq; without them this book would certainly not have been possible.

Introduction

The Kalashnikov assault rifle, generically known as the AK-47, is the most famous small arm ever produced. It is a weapon that has transcended both its designer and country of origin to become the most prolific and iconic weapon in the world. In addition, it has become a brand that has been used to sell everything from t-shirts to vodka.

Although it was introduced in the late 1940s, it first made its decisive presence felt in action during the Vietnam War, when the Vietnamese Communists were supplied it by China and the weapon's durability became legendary. Since then it has been employed in practically every single conflict across the globe. It has surely became the symbol of all wars of national liberation, but perhaps its most iconic moment came in the hands of the Mujahideen fighting to oust the Soviets from Afghanistan.

The sleek assault rifle with the banana-shaped magazine is synonymous with militant insurgency and is a common sight in all the world's trouble spots, including Afghanistan, Iraq, Libya, Somalia and Yemen. It is a very robust and easy to use weapon that has been in service throughout the world for over sixty years. The Kalashnikov's is a story of unrivalled success.

To give some measure of this weapon's success (perhaps proliferation might be a better word), an estimated fifty to one hundred million have been produced. The next most common assault rifles are the German-designed G3 and the American M16, of which fifteen and five million have been manufactured respectively. The AK easily outstrips the US-produced M16 by a factor of ten to one.

In addition, the Kalashnikov has been the inspiration behind assault rifles made in Finland, Israel and South Africa, to name but a few. Most notably the Finnish Sako M60, M62 and M76, the Israeli Galil ARM/AR assault rifles and the South African R4 all draw on the design qualities of the AK-47.

I first came across the Kalashnikov assault rifle during the 1980s when I was attending a civilian staff course at Sandhurst Military Academy, while serving with British Defence Intelligence. At the time – at the height of the Cold War – our obsession was the Soviet Union and everyone was well aware of the durability and ubiquitous nature of the famed Kalashnikov. We were taken out on to the ranges by an instructor whose role was to familiarise us with standard issue small arms of the world's armies. Among them was the American M16, the British SA80 (which had just come into service, despite numerous teething problems), the Soviet AK-47 and its slightly younger and lighter brother the AKM, and the Soviet Army's latest version, the AK-74. The latter, plus the light machine-gun variant known as the RPK-74, had only fallen into British hands in recent years courtesy of the Mujahideen in Afghanistan.

The instructor proceeded to fire a few rounds from the AKM, warning us of the dangers of muzzle lift and the technique known as 'spray and pray' so favoured by the world's guerrilla and insurgent forces. He also warned that a rate of fire of 600 rounds per minute means it's very easy to empty the thirty-round magazine. In addition, one has to be careful for some Kalashnikov variants expel their spent cartridges to the right by as much as 40ft, thereby showering everyone in the vicinity.

Then he proceeded to remove the receiver cover and drop a handful of dirt into the receiver mechanism. 'These', he said, pointing at the array of firepower lying on the ground, 'will jam every time without fail if not kept clean. This', he said, reassembling the weapon, 'makes no difference.' He handed the assault rifle to an alarmed student, saying, 'Go on then, sir, you know you want to. It's idiot proof and very, very reliable.' He gave everyone else a reassuring wink.

It was then that I fully realised that every time you see news footage of a conflict in some far-off country, it is inevitable that some of the combatants are likely to be armed with the AK-47. The Kalashnikov assault rifle long ago escaped its creator and has been manufactured by well over a dozen countries. As a result it has thwarted all attempts at effective small arms control and helped fuel almost every single conflict since the end of the Second World War. So many have been manufactured that you could give two to every man, woman and child in the UK.

In the years to come I was to witness how Chinese copies of the AK-47 helped drive the Soviets from Afghanistan and then brought the Taliban to power. I saw colleagues trapped in Beirut as local warlords' militia ran amok brandishing AK-47s. I watched as Libya shipped over AK-47s to support the Republican cause in Northern Ireland. In the Balkans I saw how Serbian-manufactured AK-47s helped bring bloody misery to a disintegrating Yugoslavia. I found myself peering down the barrels of Iraqi-produced Tabuk AK-47s in the hands of young Republican Guardsmen on the streets of Baghdad and elsewhere. In more recent years the AK-47 has cut a swathe through the populations of the Democratic Republic of Congo, Sierra Leone, Liberia and the Ivory Coast. It has been used to gun down protestors in Egypt, Libya, Syria and Yemen. The Russians continue to use its successors, the AK-74M and AK-100.

It should go without saying that this book is in no way intended to glorify the Kalashnikov. The saying goes 'guns don't kill people, people kill people', but this is nonsense: you need the tools to wage war as well as the intent. Unfortunately, the

If posting from outside of the UK please affix stamp here

Pen & Sword Books
FREEPOST SF5
47 Church Street
BARNSLEY
South Yorkshire
S70 2BR

AK-47 has proved to be a highly effective tool over the last sixty years. Rather, this book's aim is to try to explain the reasons for the weapon's longevity, its ready availability and the impact that it has had on key conflicts since the 1950s. It is also intended to provide a visual guide to the evolution of the most famous weapon in the world.

Over the years I have handled Albanian-, Bulgarian-, Chinese-, Hungarian-, North Korean-, Romanian-, Soviet- and Yugoslavian-manufactured Kalashnikovs in both assault rifle and light machine-gun configurations. The quality of the work and the weight may vary but essentially it remains the same beast. This is truly the people's gun.

The man who started it all. A young Mikhail Kalashnikov poses in his design office with colleagues. Although the AK-47 was undoubtedly a team effort, Soviet propagandists ensured that he got the credit. Never in his wildest dreams could he have imagined that he would become a household name, or that his weapon would cause such global mayhem. (*Author's Collection*)

II. ОБЛЕГЧЕННЫЙ 7,62-*мм* АВТОМАТ КАЛАШНИКОВА (АК)

Облегченный автомат (рис. 5) по устройству и баллистическим данным аналогичен необлегченному автомату, но имеет следующие особенности:

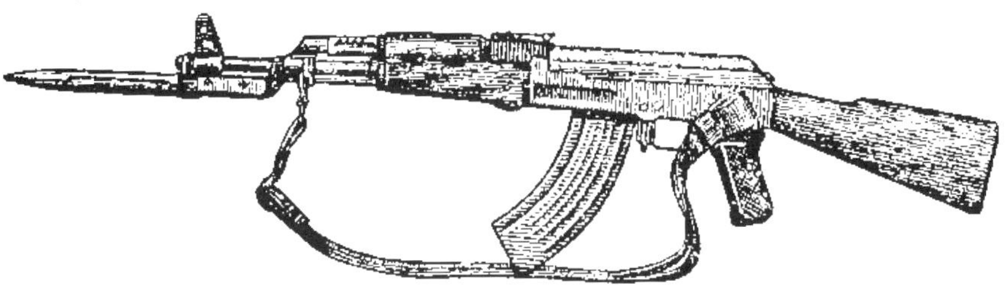

Рис. 5. Общий вид облегченного автомата

An early illustration of the 7.62 x 39mm Avtomat Kalashnikova Model 1947 or AK-47 for short – it is easily recognised by the milled receiver featuring the rectangle just above the magazine and the smooth top receiver cover. The barrel also features a round muzzle brake, rather than the slanted one that appeared on the later AKM. (*Author's Collection*)

Chapter One

The People's Gun

As early as 1916 the Russians developed an Avtomat (automatic) that was neither a rifle nor a submachine gun, but it was not very successful. Three years later Mikhail Timofeyevich Kalashnikov, the man who was to develop the most enduring assault rifle of all time, was born in Kurya. Just before the outbreak of the Second World War he was drafted into the Red Army's armoured corps as a driver/mechanic. Following Adolf Hitler's invasion of the Soviet Union in 1941, Kalashnikov was wounded in an engagement near Bryansk, gaining the Order of the Red Star for gallantry. Judging by how severely Joseph Stalin's tanks were mauled by Hitler's panzers that summer, Kalashnikov was extremely lucky to escape with his life.

It was a Nazi atrocity that led to the wounded Kalashnikov developing the world famous assault rifle that continues to bears his name even today. German troops, possibly SS, armed with submachine guns massacred some of his wounded comrades and this made him determined to design an effective automatic infantry small arm that would give them greater firepower.

What emerged was a weapon that is robust and easy to operate, with minimal working parts. It can be fired in single shot or automatic fire modes by depressing the lever located near the trigger mechanism. With the trigger held depressed in automatic it can pump out 600 rounds in a minute – however, as the magazine only holds 30 rounds, these will be gone in just over three seconds. Its official maximum range is 800–1,000 metres, but in fact even in trained hands it is only accurate to about half that distance.

Both Russia and Germany were very slow to develop a weapon that combined rapid fire with a rifle's accuracy in the critical mid-zone of combat, and the existing rifles and submachine guns did not really fulfil this requirement. Most rifles were calibrated for 1,000 metres plus, while submachine guns and machine pistols were good for close quarters; what was needed was something in the 250–750 metre range. Such a weapon would lessen the infantry's need for heavy and light machine guns.

Certainly the Red Army made greater use of the submachine gun than did any other army during the Second World War. This was in part due to many of their factories being overrun and they needed something simple to produce. Finnish small arms designer Aimo J. Lahti produced his first submachine gun in 1922, but his main model was the 1931 Suomi, which utilised a drum magazine. The Suomi saw action in the Russo–Finnish Winter War of 1939/40. Soviet designer Vasily A. Degtyarev copied Lahti's magazine and details from Schmeisser's MP28 to produce the Pistolet Pulomet (Machine Pistol) Degtyarev 1934 or PPD34 for short. This was followed by the PPD40.

One of the most famous Soviet designers, Georgi S. Shpagin, Hero of Socialist Labour and later a lieutenant general in the Red Army, then produced one of the most iconic submachine guns of the Second World War. After the Winter War he developed a simpler version of the PPD34/38, redesigning it as the 7.62mm Pistolet Pulomet Shpagin 1941, better known as the PPSh-41. It can fire in automatic or semi-automatic modes.

Although the PPSh-41 did not enter service until mid-1942, by 1945 five million had been manufactured and, like Kalashnikov's AK-47, it became synonymous with post-war Communist military and guerrilla forces around the world. Despite being obsolete after the Second World War, the PPSh-41 was produced in China as the Type 50, in Hungary as the 48M, in North Korea as the Type 49 and in North Vietnam as the K-50M.

Alexei Sudarev also developed the PPS-42 in 1941, followed by the PPS-43, which were both much simpler than the PPSh-41 and featured a folding metal stock that went up over the weapon's body. The PPS-43 was subsequently produced by China as the Type 43 and, along with the Type 50 and K-50, it became a very popular weapon with Asian guerrilla armies.

In the meantime, while still on sick leave, Kalashnikov produced a prototype weapon chambered for the standard Soviet 7.62mm pistol cartridge during 1942. However, when the PPS-43 Pistolet-Pulomet Shpagin was tested, it was adopted instead. Kalashnikov's efforts did not go unnoticed, though, and with the help of General Anatoly Blagonravov, a leading Soviet arms expert, he became a small arms designer. For the rest of the war Kalashnikov spent most of his time improving and fine-tuning existing weapons designs. At the end of the Second World War in 1945 he returned to his design for a submachine gun using the 7.62mm x 39mm M1943 round, incorporating a locked breech and a gas-operated mechanism.

It is quite likely that he borrowed ideas from German wartime assault rifles. During the 1930s Germany had experimented with automatic rifles, leading to a series of weapons produced during the Second World War. The best known of these was the MP43/StG44 Sturmgewehr or assault rifle designed by Hugo Schmeisser. This was constructed from pressed and diecast components with a wooden or plastic shoulder stock. It featured a forward-curving box magazine for the thirty 7.92mm x 57mm short calibre rounds and had an effective range of 800 metres.

The StG44 with its gas-operated system could be fired on fully automatic as well as single shot by pressing a selector button above the trigger. This weapon enabled each infantryman issued with it to become his own light machine-gunner. Just as importantly, it proved to be robust and reliable. This superb if rather bulky assault rifle appeared rather late in the war and, although cheap to produce (as it relied on fabricated rather than machined components), German weapons manufacturers struggled to meet demand in the face of dwindling raw materials and Allied bomber attacks.

In the closing days of the war a cheap and cheerful version known as the Volksturm Gewehr or people's assault rifle was also produced. It could only be fired in single shot mode and with a range of just 300 metres was not terribly useful. The Germans also developed the lightweight Fallschirmjäger Gewehr 42 for its airborne forces but this was only produced in small numbers. Over the years there has been much heated debate about the influence the StG44 had on Kalashnikov – suffice it to say that the Soviets examined all captured German weaponry and would have been well aware of the StG44's merits.

When the Soviet Artillery Commission tested Kalashnikov's final prototype they found it easy to use and reliable; it had an effective range of 400 metres, fired 600 rounds per minute from a thirty-round magazine and weighed 3kg (the StG44 weighs in at 5.22kg). In single shot mode it has hardly any recoil, but in automatic mode it does tend to suffer from muzzle climb, as can be seen from news footage from around the world. The Artillery Commission decided that the Red Army should adopt this weapon as the Automat Kalaschnikova 7.62mm Model 1947, abbreviated to AK-47. Despite its designation, it did not in fact enter service until 1949–51. Along with the 7.62mm Samosarjadvij Karabin Simonowa SKS45 carbine and the 7.62mm Roschnoi Pulomet Degtjarewa RPD light machine gun, the Kalashnikov became the mainstay of Soviet infantry weaponry, and by the end of the 1950s the AK-47 had replaced the SKS as the standard Soviet infantry weapon. Notably in 1959 the Kalashnikov Modificatsonni or AKM appeared. It was lighter, with a receiver stamped out of sheet steel instead of milled; it also has a distinctive ribbed receiver cover, dimples above the magazine well and ridges on the forehand guard. The AKM went into production the following year and shortly afterwards other countries began to produce their own versions. The Chinese, Egyptians, Iraqis, Palestinian Liberation Organisation and even the Afghan Mujahideen produced cloned copies of the AKM.

The Ruchnoi Pulomet Kalashnikova or RPK, the light machine-gun variant of the Kalashnikov, appeared in the 1960s. This is a strengthened AKM with a longer and heavier barrel. The Kalashnikov design teams also came up with the Pulomet Kalashnikova, which served as the Soviets' medium machine gun using a 7.62 x

54mm round. This was followed by a modernised version known as the PKM. Both the RPK and PKM have been widely copied.

China got round licencing niceties by manufacturing the Chinese Communist or Chicom Type 56, which was a direct copy of the AK-47 but with a folding triangular bayonet. Indeed, the Chinese went on to copy just about every single Soviet small arm design. Perhaps not surprisingly, the AK-47 first saw combat with Communist guerrillas. The Communist Malayan Races Liberation Army deployed it in very small numbers during the Malayan Emergency of 1948–60. The Egyptian Army used the Kalashnikov against the British, French and Israelis during the Suez crisis of 1956. Also that year the Soviet Army used it in Budapest to help suppress the Hungarian uprising.

The Israelis captured considerable quantities of them and continued to do so during the Arab–Israeli conflicts of 1948–82. Large stocks were also seized from the Palestinian Liberation Organisation in southern Lebanon. The Israelis were so impressed with it that for a time it was the weapon of choice of the Israeli Defence Forces' elite formations, such as the Golani Commandos, who were issued with AK-47s and AKMs, and the Kommando Yami (Naval Commandos), who used the AK-47.

The ubiquitous Kalashnikov had become the workhorse of the world by 1970, being readily available almost anywhere. Notably it saw service in Africa, with the Zimbabwe African People's Union or ZAPU and the Zimbabwe African National Union or ZANU in Rhodesia, with the Western Somali Liberation Front or WSLF in the disputed Ogaden, and in Eritrea with the Eritrean Liberation Front.

Fortunately large numbers of Kalashnikovs were kept out of Northern Ireland. Some 250 AKMs were seized off the coast of County Waterford in 1973, and two years later 106 Simonovs and Kalashnikovs were seized at Schiphol Airport in the Netherlands, all of which had been destined for the Irish Republican Army.

By the early 1980s the Kalashnikov was, ironically, being used against the Soviet Army by the Afghan Mujahideen in their guerrilla war against the Russian-supported Marxist Afghan government. During the Iran–Iraq War both the Iraqi Army and the Iranian Revolutionary Guards fought with it. From the outset President Saddam Hussein's Special Forces commandos carried out running street battles armed with the AK-47 in the contested Iranian city of Khorramshahr (formerly Muhammara).

(Opposite, below) Russian troops at the entrance to a subway station in Berlin armed again with the PPSh-41. This weapon was subsequently produced by China as the Type 50, by Hungary as the 48M and by North Korea as the Type 49. Mikhail Kalashnikov's tinkering with submachine guns led first to a carbine design and then to an assault rifle design that became the AK-47. (*Author's Collection*)

The famous PPSh-41 submachine gun designed by George S. Shpagin was widely used by the Russian armies throughout the Second World War. The Russians decided they also needed a weapon that combined the rate of fire of a submachine gun and the range of a rifle. (*Author's Collection*)

Polish school children undergoing weapons training in the 1950s with the PPSh-41 submachine gun, which continued to see action right up until the 1980s in the hands of the Afghan Mujahideen. (*Author's Collection*)

The same group of children examine the much newer AK-47. Despite its official classification as the Kalashnikov assault rifle Model 1947, the Soviet Army did not adopt it until two years later. Early Type 1 receiver models had stamped sheet metal receivers, but quality control problems resulted in these being replaced by the easier-to-produce but heavier milled receivers of the Types 2 and 3. The upshot was that large numbers of the AK-47 were not available until the mid-1950s. (*Author's Collection*)

A Soviet soldier takes his oath of loyalty to the Motherland while clasping his AK-47. Use of this weapon rapidly spread to the East European armies of the Warsaw Pact and around the world. Its simplicity also meant lots of countries made their own versions. (*Author's Collection*)

This is the slightly later Avtomat Kalashnikova Modernizirovanniy or AKM – the modernised Kalashnikov. This used a much lighter stamped sheet metal receiver, identifiable by the recessed dimple above the magazine, and featured a slanted muzzle brake to compensate for barrel lift from the recoil. Although all Kalashnikovs are generically known as AK-47s, the AKM is the most common and most copied type in existence. (*Author's Collection*)

This Romanian soldier is firing the Ruchnoi Pulomet Kalashnikova or RPK, the light machine-gun variant of the Kalashnikov which appeared in the 1960s. This is a strengthened AKM with a longer and heavier barrel. The RPK is used and produced throughout the world. This particular weapon is probably the Romanian-manufactured Puşcă Mitralieră Model 1964. (*Lieutenant-Colonel Dragoş Anghelache*)

A Chinese sailor with the Chinese Type 56, easily identifiable by its distinctive fold-out bayonet and lack of muzzle brake. This weapon is almost as common as the AKM and has been supplied to many countries around the world since 1956. The Type 56-1 featured the same folding metal stock as the AKM, but typically lacked the folding bayonet. (*US DoD/PH1 Charles Mussi*)

Soviet Naval Infantry posing with the AK-74 developed in the early 1970s as a 5.45 x 39mm assault rifle. It first saw action with the Soviet forces in Afghanistan. Although unlicensed versions were also produced by Bulgaria, China, the former East Germany and Romania, only around five million have been made so it remains easily eclipsed by the AK-47/AKM. (*Author's Collection*)

Soviet infantry on parade with the AK-74 at the height of the Cold War. The AK-74 has a very large muzzle brake, a different shoulder pad to the AKM (rubber and serrated), and the stock features a groove that not only lightens it but also prevents the accidental loading of AKM ammunition. (*Author's Collection*)

A shortened carbine variant of the AKS-74 (S = Skladnoy, or folding stock) known as the AKS-74U (U = Ukorochenniy, or shortened) was issued in the late 1970s to armoured vehicle crews and Special Forces. It was a favourite with the late al-Qaeda leader Osama bin Laden, who liked to be photographed with the weapon. The AKS-74U, like the AK-74, received its baptism of fire in Afghanistan with the Soviet Army. *(Roman Stepanov)*

Palestinian Liberation Organisation fighters armed with the AK-47 and the Roschnoi Pulomet Degtjarewa RPD 7.62mm light machine gun (forerunner of the RPK). By the 1960s the AK-47 had become the preferred weapon of most guerrilla armies around the world. *(Author's Collection)*

It was in the hands of the Mujahideen or 'Soldiers of God' fighting the Soviet-backed government in Afghanistan in the 1980s that the Kalashnikov first really captured the public imagination. These fighters are armed with AKMs, while the man on the left has the RPG-7 (rocket-propelled grenade), another ubiquitous Soviet-designed weapon. (*Author's Collection*)

Two Soviet prisoners under the watchful eye of the Mujahideen. Their guards are equipped with AKMs and the man on the left is clearly taking no chances, having fixed his bayonet. (*Julian Gearing*)

A Soviet armoured vehicle crewman in Afghanistan posing with his AKS-74, the folding stock variant of the AK-74. Unlike the under-folding stock of the AKS-47 and the AKMS, this has a triangular stock that folds to the left of the weapon. (*Igor Bondarets*)

An Afghan security guard firing his AKM at Forward Operating Base Torkham, Nangarhar Province, in May 2009. The Kalashnikov remains in widespread use with both the Afghan security forces, the Taliban and other Afghan insurgent groups. (*US DoD/Sgt Mathew C. Moeller*)

Iraqi police under instruction by American trainers from the US 7th Cavalry Regiment, 1st Cavalry Division. The policeman is holding a very tatty-looking AKM, while the instructor has the folding stock AKMS. These men are clearly posing for the camera conducting blank firing drills as neither weapon has a magazine. (*US DoD/Senior Airman Steve Czyz*)

An Afghan National Army soldier with the Romanian PM md 65 folding stock AKMS, which has a distinctive pistol fore grip; this does not curve forward as on the earlier AIM or PM md 63 (the Romanian version of the Soviet AKM) and has a rounded rather than squared-off end. Very large numbers of Romanian-produced assault rifles have been exported to developing countries around the world and it is regularly seen in the world's troublespots. (*US DoD/Sgt Sean A. Conley*)

Afghan Border Police on patrol from Combat Outpost Margah, Paktika Province, April 2010. While the man on the left has the standard AKM, his colleague is holding the Egyptian Maadi Misr, clearly identifiable by the single stem wire stock and triangular butt stop. Egyptian-produced Kalashnikov clones have been sold around the world. (*US DoD/Sgt Derec Pierson*)

Another Afghan policeman, this time armed with a Hungarian AMD-65 – a short version of the Hungarian AKM-63 AKM. It has a single strut folding stock, and the barrel has been shortened and given a special muzzle brake. It has a plastic pistol grip and a forward pistol grip under the fore grip, which has a ventilated metal hand guard. (*MoD/UK Maj Paul Smyth*)

Another photo of the Hungarian-produced Kalashnikov, which is very popular with the Afghan National Police. This officer is training female recruits at the Kabul Military Training Centre. (*US DoD/Senior Airman Matt Davis*)

This Iraqi soldier is using a later export model of the Chinese Type 56 known as the Type 56-2. This assault rifle's stock is made of sheet metal with a plastic cheek piece, which is clearly visible in this photo. The drawback with this is that the stock folds to the right and obstructs the fire selector above the trigger. These rifles generally have no fitted bayonet and normally have red-brown furniture made of layers of fibreglass soaked in epoxy plastic. (*US DoD/Staff Sgt Dallas Edwards*)

Iraqi Police with battered AKMs, which have clearly seen a lot of action over the years. These men are in Taji undergoing the Iraqi Police Prep course in October 2007 under the watchful eyes of members of the US 7th Cavalry Regiment. The lack of magazines in their weapons indicates they are conducting basic firearms drills. (*US DoD/Senior Airman Steve Czyz*)

By 2010 the Afghan National Army was still reliant on the Kalashnikov as its main small arm in the war against Taliban insurgents. This young soldier is equipped with an AKM; note that the fire selector above the trigger is set to off (pushing it down selects semi- or automatic fire), so he is not expecting any trouble. (*US DoD*)

Chapter Two

Freedom Fighters' Choice

The uneducated peasant soldiers and militias of Africa soon learned to appreciate the simplicity and ruggedness of the AK-47. Across North Africa and sub-Saharan Africa barely trained men were able to pick up the Kalashnikov and wage war with ease. In rainforest and desert alike the weapon was regularly abused by ill-trained troops who had little time or inclination for the formalities of weapon maintenance – and yet the AK-47 did its job time and time again.

Unfortunately it was also used to arm Africa's boy soldiers from Liberia to Uganda – children snatched from their parents and drafted into ill-disciplined, brutal and bloodthirsty militias. Teenage hormones, abusive discipline and a ready supply of assault rifles were the deadly ingredients for a savage recipe for innumerable atrocities and massacres, which were often given the facile veneer of some political or sectarian cause.

From the Horn of Africa to Central America the AK-47 became ubiquitous. Soviet military aid was seen as a way of covert power projection, and was also employed to create military dependency by assisting countries lacking economic and technical ability. This was best exemplified by Moscow's Kalashnikov diplomacy – make them grateful, make them dependent. Throughout the Cold War Moscow's military aid to its Communist allies came at a high price. Wherever there were Soviet weapons, there were invariably Soviet advisers and technicians. It also meant that for decades the Soviet Union and the United States of America were able to fight a series of very bloody proxy wars without ever coming directly to blows.

At the height of the Cold War in the 1980s the Soviet Union exported billions of dollars' worth of arms to numerous developing world countries. Intelligence analysts watched with a mixture of dismay and awe as cargo ship after cargo ship sailed from Nikolayev in Ukraine stacked to the gunnels with weapons bound for ports such as Assab (Ethiopia), Luanda (Angola), Tartus (Syria) and Tripoli (Libya). Much of this equipment came out of Russian strategic reserves and was either very old or had been superseded by a newer model.

From the 1950s onwards guerrilla armies seeking to throw off their colonial rulers in Africa began to receive Soviet and Chinese weapons. The AK-47 and Type 56 soon became commonplace as Moscow and Beijing vied for influence in post-colonial Africa. By 1980 Moscow had supplied weapons to almost half the states on the African continent, which had been blighted by a succession of unending wars.

North Korea produced vastly more Kalashnikovs, known as the Type 58 and Type 68, than it ever needed and exports were made to Africa and the Middle East. Similarly, large numbers of the Hungarian AKM-63 and AMD-65 and the Romanian AIM and AKMS were exported to the armies of the developing world.

In addition, many African countries have made their own copies of the Kalashnikov. Ethiopia's state-run Gafat Armament Engineering Complex manufactures the Et-97. In recent years Nigeria had plans to mass-produce a version of the AK-47 known as the OBJ-006, and Sudan produces a copy of the Chinese Type 56 known as the MAZ.

During the 1960s China extended military aid to Uganda, Zanzibar/Tanganyika (Tanzania) and the Belgian Congo. Chinese guns went to the Stanleyville regime during the Congo Civil War, to the Watusis in the Rwanda–Burundi conflict, and to the nationalist movement in Niger. China also supplied Guinea, Sierra Leone, Sudan and Zaire.

The Belgian Congo gained independence in 1960 and in the civil war that followed Sino–Soviet rivalry was evident in their support for the rival protagonists. China backed the bloodthirsty Simbas against the Congolese Army. Soviet weapons subsequently proved vital to the campaigns fought against the Portuguese by UNITA (União Nacional para a Independência Total de Angola in Angola), by PAIGG (Partido Africano de Independencia de Guine e Cabo Verde) in Guinea-Bissau and by FRELIMO (Frente de Libertacoa de Mocambique) in Mozambique.

In the case of the struggle for black majority rule in Rhodesia (modern-day Zimbabwe), the Soviets backed ZIPRA (Zimbabwe People's Revolutionary Army) and the Chinese backed ZANLA (Zimbabwe African Liberation Army). China also sent military aid to rebels in Angola, Guinea-Bissau, Mozambique, Rhodesia and South Africa. As a result the Chinese Type 56 and SKS became as commonplace in Africa as the AK-47.

Beijing, though, could never really compete with Moscow. Once African revolutionaries were in power Moscow helped them stay there, supplying weapons to counter internal insurgencies. This was most notable in Angola, Ethiopia and Mozambique; in the case of Angola, China backed the wrong faction.

Soviet arms also fuelled the conflicts in Ethiopia and Somalia, especially the Ogaden War in 1977 and the Eritrea–Ethiopia War. At one point Ethiopia accounted for half of Soviet arms sales to sub-Saharan Africa. In the mid-1980s Moscow increased its support to counter-insurgency efforts in Angola, Ethiopia and Mozambique, which amounted to billions of dollars' worth of weaponry including tens of thousands of assault rifles. In the meantime some African countries such as Sudan became key customers for Chinese weapons, including the Type 56.

Ironically though, many Soviet client states ended up becoming liabilities rather than assets. In the long term, supplying the Kalashnikov and other weapons to Africa did not buy Moscow any lasting influence. Somalia went against Soviet wishes over the Ogaden, as did Ethiopia over its handling of Eritrea. The Soviet-backed Angolan and Mozambican governments proved incapable of defeating the guerrillas trying to overthrow them. Indeed, none of the Marxist governments in Angola, Mozambique and Ethiopia was able to defeat the rebels, leaving their economies in tatters.

Angola and Ethiopia were Moscow's most important regional allies and customers. In the late 1970s, when Ethiopia was at war with Somalia, the Soviets delivered them weapons. Ironically, Somalia had been a former Soviet ally and had received weapons before the two countries had fallen out. In the case of Angola, in 1986 Soviet arms supplies were to the value of $2 billion. Two years later, when Cuban forces propping up the Marxist Angolan government were in the process of withdrawing, they handed over Soviet weaponry to government forces. Likewise Mozambique, also blighted by civil war, spent $1 billion on weapons.

In the midst of a bitter and bloody civil war Ethiopia also took receipt of Soviet arms in the mid-1980s. By the end of the decade it had spent over $5.4 billion on weapons while its population starved to death. Equipment delivered to the port of Assab for the 1985 campaign in Eritrea consisted of small arms, tanks and armoured personnel carriers. Three years later the writing was on the wall for Ethiopia's Marxist government after its forces suffered defeats at Afabet, Keren and Inda Selassie. Libya used Soviet arms against Chad in 1983–88 and lost $1 billion worth of arms in 1987 at Quadi Doum and at Maartan-as-sarra. Such losses seemed to matter very little.

One country where the Kalashnikov has had a particularly devastating effect because of its ready availability is the Democratic Republic of Congo (DRC). 'Blood' diamonds and other looted natural resources such as Coltan, copper, gold and timber have fuelled the bloody conflicts there. Such revenues helped arm and equip the many feuding militia armies in the region. AK-47s can be purchased in the Congo for $50–100 a gun, and bulk orders obviously attract a considerable discount. The vast majority of the weapons seized by the UN in the DRC are Chinese Type 56 and Type 56-1. This is because Chinese weapons are cheaper than those from Russia, Ukraine or other Eastern European countries. China of course does not deliberately infringe UN sanctions that prohibit gun sales to Congo – arms dealers and middlemen do that.

The Second Congo War, which ran from 1998 to 2003 and involved seven foreign armies, was so devastating that it is sometime called the African World War. The subsequent presence of UN peacekeepers in the east of the country did little to stop the bloodletting. The DRC war erupted when Rwanda, Uganda and later Burundi-backed Congolese rebels sought to oust President Laurent Kabila. Troops from Zimbabwe, Namibia and Angola stepped in to support the Congolese government.

Under Operation Sovereign Legitimacy Zimbabwe deployed 12,000 troops, tanks and fighter aircraft to the DRC's resource-rich Kasai Oriental and Katanga provinces to help prop up the government. Zimbabwe had already provided a steady supply of stockpiled North Korean weapons to Kabila's rebels during the First Congo War before he came to power. These would have included Type 68 and Type 56 assault rifles.

While it may have been the freedom fighters' weapon of choice, rather than liberation the AK-47 simply brought bloodshed and misery to the vast continent of Africa.

The Kalashnikov has been used in most of the wars fought in Africa since the Second World War, most notably in Angola, Chad, Ethiopia, Mozambique, Namibia and Somalia. During the 1970s and 1980s Ethiopia and Eritrea were the scene of a particularly protracted and bloody civil war. This member of the Ethiopian National Defence Force is firing an AK-47 with a milled receiver, though the ribbed receiver cover is clearly from an AKM – many guns end up a mish-mash of different parts from different guns. (*US DoD/Eric A. Clement*)

One of the most common types of Kalashnikov in Africa is the Chinese Type 56; here a soldier from the Ugandan People's Defence Force is fielding an example. Uganda has endured many wars over the years, and the Ugandan Army has struggled to contain the Lord's Resistance Army which operates in the north of the country. (*US DoD/Master Sgt Ruby Zarzyczny*)

Another member of the UPDF equipped with the Type 56. Like all Kalashnikov variants it is very robust and easy to use. Uganda has around 40,000 men under arms. (*US DoD/Master Sgt Ruby Zarzyczny*)

In 1969 Angolan guerrillas fought the Portuguese Army for control of the country. During the opening stages of the wars of liberation against the colonial powers in Africa, the various guerrilla armies were very poorly armed. Once China and the Soviet Union stepped in the situation very quickly changed. (*Author's Collection*)

This guerrilla, photographed in 1968, is equipped with the Soviet SKS carbine, which was very much an interim weapon while problems with the AK-47 were ironed out. Like the Chinese Type 56, it is had a fold-out bayonet and was copied by many countries. (*Author's Collection*)

These female fighters are armed with a variety of firearms; the two in the middle are holding Soviet PPS-43 submachine guns, a type which was essentially a cheap and cheerful version of the PPSh-41. Both types were exported after the Second World War. The women on the left are holding a Czech Model 24/26 submachine gun that became the inspiration for the Israeli Uzi. (*Author's Collection*)

These Guinean rebels started out in the late 1960s with outdated weapons but soon found themselves equipped with modern Czech, Soviet, Chinese and American arms. This group of fighters are sporting various types of rocket-propelled grenades probably supplied by the Chinese. (*Author's Collection*)

During the guerrilla war in what is now Zimbabwe, fought during the 1960s and 1970s, the Rhodesian Defence Force did all it could to contain the guerrilla armies fighting to end white minority rule. These men are searching a guerrilla base for weapons and ammunition. (*Author's Collection*)

Ugandan soldiers with AKM assault rifles. (*US DoD/Tech Sgt Jeremy T. Lock*)

The state of Somalia has been blighted by civil war and piracy. It has become a haven for various terrorist groups and is a key source of illegal weapons in the Horn of Africa. This militiaman, photographed in 1992, is keeping order with an AKM. The following year Somali warlords humbled the US military by shooting down a Black Hawk helicopter in Mogadishu using RPGs and AK-47s. (*Via Author*)

Across the Gulf of Aden Yemen has been plagued by al-Qaeda and the Houthi rebellion. The Yemeni soldier on the left is equipped with a Romanian AKMS. (*Via Author*)

Niger has experienced problems with a terror group known as al-Qaeda in the Islamic Mahgreb and neighbouring Libya. These Niger troops from the 322nd Parachute Regiment are on exercise with folding-stock AKMS. Niger has long struggled to secure its borders with an army of only 5,200 men, supported by a similar number of paramilitaries. (*US DoD*)

This Ethiopian tribesman has a rather ancient AK-47 slung over his shoulder. Note the homemade strap made from a piece of rag. (*Via Author*)

A member of the 3rd US Infantry Regiment instructing soldiers from the Ugandan People's Defence Force in 2008. Like many African countries, Uganda was enlisted in the war against terror, especially after the bomb attacks on the US embassies in Kenya and Tanzania a decade earlier. These Ugandan soldiers are armed with the Chinese Type 56. (*US DoD*)

Another Ugandan soldier with a Type 56 under instruction. (*US DoD/Master Sgt Jeremiah Erickson*)

Liberia, Ivory Coast and Sierra Leone have all been plagued by civil wars that were fuelled by the ready availability of Kalashnikov assault rifles. These Liberian National Guardsmen at the Edward Binyah Kesselly AK-47 rifle range are being mentored by a US Army sergeant in April 2009. Their weapon of choice is the AKM. The Liberian Army numbers around 10,000 men. (*US DoD*)

This Ethiopian warrior is carrying what looks to be a Bulgarian AK-47, judging by the milled receiver and the plum-coloured plastic furniture. Tribal warfare has long been endemic along Ethiopia's borders with Eritrea, Kenya, Somalia and Sudan. Cattle rustling and gun running have also long been a way of life in this part of the world. (*Via Author*)

Ugandans on the counter-terrorism training course run by the Combined Joint Task Force–Horn of Africa at the Kasenyi Military Training Centre in 2009. (*US DoD/Master Sgt Ruby Zarzyczny*)

Soldiers of the UPDF practise medical procedures during a squad completion at Forward Operating Location Kasenyi in Uganda in April 2008. The US instructor is from the US 3rd Infantry Regiment. The 'casualty's' weapon is a Type 56. (*US DoD/Tech Sgt Jeremy T. Lock*)

More Ugandan recruits armed with Kalashnikovs undergoing training at Kasenyi in early 2008. (*US DoD/Tech Sgt Jeremy T. Lock*)

A Beninese soldier with the ubiquitous Chinese Type 56 on a joint US training exercise in June 2009. Benin's army is tiny, numbering fewer than 4,500 men, but, like many African countries, it is equipped with Chinese small arms. (*US DoD/Lance Cpl Jad Sleiman*)

Chapter Three

Viet Cong to Mujahideen

It was in the dense jungles of Southeast Asia that the superiority of the robust Kalashnikov design first became really apparent. The difference in the design philosophy resulted in it performing much better than Western weapons in the majority of roles during the Vietnam War. The early version of the American M16 assault rifle deployed in Vietnam created many problems for US troops. Most importantly, the unpredictable performance of the M16 was characterised as 'confidence sapping' – the reliability of a soldier's weapon is the last thing he should be worrying about. While the M16 performed very well in normal conditions, it began to malfunction when battle-tested in the humid jungle terrain of Vietnam.

The AK-47 was not in widespread use by 1950 and did not see combat during the Korean War. Instead the Communist North Korean People's Army was equipped with the Soviet PPSh-41 submachine gun and SKS carbine. North Korea subsequently produced its own copies of the AK-47, known as the Type 58 (AK-47) and Type 68 (AKM), as well as the Type 88 (AK-74) at the Kanggye Industrial Zone in the north of the country.

The Type 58 has a milled receiver, making it, in common with other AK-47 copies, a rather heavy weapon and particularly heavy on the fore grip. North Korea's Type 68 is identical to the Soviet AKM except that the front grip is closer to that of the AK-47 than on the thicker AKM style. It also produced the SKS as the Type 63. During the 1980s North Korea was exporting up to $500 million worth of weapons a year. However, its key customers Iran and Zimbabwe never seemed very happy with what was delivered and North Korea appears to have been used as a last resort when other suppliers were not available.

Soviet and Chinese Kalashnikovs first really came into public consciousness on the battlefields of Vietnam. Another weapon that made its mark there was the rocket-propelled grenade (RPG). The infamous Soviet RPG-7 anti-tank weapon capable of penetrating 10 inches of armour made its debut in Vietnam. Today it is the most common anti-tank weapon in the world. The Chinese predictably copied this weapon, designating it the Type 69.

Initially the Communist forces in Vietnam relied upon captured French weaponry.

Although some standardisation was achieved in the later stages of the war, there was still a huge amount of variety in Communist small arms. The one weapon that was standardised was the AK-47 in its many guises. Soviet, Chinese and Czech (Model 58) versions were all supplied in quantity. The Soviet SKS or Chicom Type 56 semi-automatic rifle was also widely used by the North Vietnamese Army (NVA) and later by the Viet Cong. China supplied the Communists with small arms while the Soviet Union provided support weapons such as rocket-propelled grenades, artillery and missiles. By 1964 Soviet and Chinese aid to North Vietnam amounted to $800 million, most of which had been military equipment, the majority of it Soviet.

Communist China was never able to compete on anything like the scale of the USSR's heavy weapons supplies. Its derivative of the Soviet T-54 tank, the Type 59, went into production in the late 1950s. During the Vietnam War the North Vietnamese were supplied with 700 Type 59 and Type 62 light tanks and Type 63 amphibious tanks (a copy of the Soviet PT-76). However, it was the Chinese-made Kalashnikov that had the most impact.

By 1968 most Viet Cong and NVA units were predominantly equipped with the Type 56 (fitted with a triangular bayonet) and Type 56-1 assault rifle – the Chinese copy of the AK-47. They also used the Type 50, the Chinese copy of the Soviet PPSh-41. The VC also produced their own version, known as the K-50M, in jungle workshops. Subsequently these weapons were also supplied to the Khmer Rouge guerrillas in Cambodia and the Pathet Lao in Laos to fight government forces.

Sufficient quantities of Kalashnikovs were received by the North for them to be shipped clandestinely to South Vietnam and used to equip most of the Viet Cong guerrilla units. Weapon supplies were shifted down the Ho Chi Minh Trail by truck and porters using bicycles. Despite the deployment of massive American air power and a sustained bombing campaign, the trail was never completely cut.

The Viet Cong in their black pyjamas, rubber sandals and straw coolie hats soon learned to respect their AK-47 and Type 56 assault rifles. From the humid jungles to the wet highlands by way of the country's muddy paddy fields, the VC and NVA found that the Kalashnikov suffered very few stoppages. While the Americans had to constantly maintain their M16s the Communists were able to carry on pressing home their attacks regardless of how dirty their weapons became. When the annual monsoons arrived the rains drenched them and covered everything in sticky mud – but it made no difference to the AK-47.

In contrast, the American GIs in Vietnam often despaired of their M16 assault rifles, which had replaced the M14 in 1965. Initially the lack of cleaning kits, a powder that fouled the barrel and the lack of chrome linings for the barrel all exacerbated the notorious unreliability of the weapon. A key problem was the spent cartridge failing to expel from the chamber after the bullet had been fired. This left the weapon inoperable in combat situations, and this resulted in deaths as the soldiers struggled to clear their rifles. Attempts to solve these problems led to the M16A1, and stoppages were greatly reduced. The subsequent M4 was introduced in 1994 and, like its predecessor, has to be properly maintained or it can suffer jams in dusty conditions.

The AK-47 is intended to act essentially as a machine gun in automatic mode first and in semi-automatic or single fire mode second. The American AR-15, renamed the M16 by the US Army, was primarily designed as a rifle first and a machine gun second. In addition, the Kalashnikov's 7.62mm calibre bullet was much larger than the M16's 5.56mm round, giving it greater stopping power and range. The AK-47 won hands down when it came to firepower. A single round from a Kalashnikov could shatter a breeze or cinder block, but an M16 would only punch a hole in it. Tests revealed that at a range of 200 yards the M16 was very accurate and in semi-automatic mode it scored five out of five hits; the AK-47, however, fared terribly at this range. After five shots were fired in semi-automatic mode, the AK-47 had scored no hits, although at a much shorter range for close-quarter combat the AK-47 performed better, especially in automatic mode.

Likewise the GIs did not fare very well with the M60 machine gun that came into service in the late 1950s. In the jungles of Southeast Asia it proved to be too heavy and, like the M16, was prone to jam at inopportune moments. Despite being a combat weapon, the M60 did not like dirt or mud. Also, if reassembled incorrectly, it would continue firing even when the operator's finger was taken off the trigger. The location of the barrel latch meant that it could snag in webbing, causing the barrel to fall out, and so the list of problems went on. Up against the Soviet PK machine gun that entered service in the mid-1960s, the M60 was found to be severely wanting. Since then it has undergone many modifications. The later American M240 and M249 machine guns proved to be much better weapons.

During the 1980s the CIA's actions in Afghanistan greatly stimulated the massive growth of the illegal arms market and this period was perhaps the heyday of the AK-47. Covert aid provided by Washington to the Mujahideen expanded from $35 million in 1982 to a staggering $600 million in 1987. On top of that, Saudi Arabia was reportedly matching American funding, which meant towards the end of the period some $1 billion a year was pouring into the Mujahideen's coffers.

Initially the CIA purchased Chinese and Egyptian Soviet-pattern guns because it was easier to obscure Washington's top-secret assistance, on the grounds that the Mujahideen's weapons could have been taken from the Afghan or Soviet armies. To the man on the street there was no way of telling whether the Mujahideen's AK-47 look-alike assault rifle had been manufactured in Bulgaria (Bulgarian Arsenal AKKM),

China (Norinco Type 56), Egypt (Maadi Misr), Poland (Lucznik Karabinek AKM), Romania (Romanian State Arsenal PM md 63/AIM), the Soviet Union (Tula or Izhevsk/Izhmash AKM) or Yugoslavia (Zastava M70).

In the early 1980s President Anwar Sadat of Egypt decided to deliver weapons, especially anti-tank and anti-aircraft missiles, to the rebels via Pakistan. Crucially Egypt was producing the Soviet man-portable SA-7 surface-to-air missile and the RPG-7. China also built both types of weapon that also ended up in militants' hands. During the early 1980s Egypt was exporting up to $1 billion worth of arms a year; while much of this went to Iraq to support its war against Iran, considerable quantities also ended up in Pakistan and Afghanistan.

At the behest of the CIA, Egypt even set up an AK-47 production line to supply the Afghan resistance movement. Once the Egyptians started producing assault rifles, the black market price for an AK-47 tumbled; when the Chinese entered the fray it fell even more and it became a buyer's market. Chinese arms soon supplemented initial supplies of ancient British .303 rifles. The Chinese proved to be reliable suppliers; they were cheap and their equipment was of a good quality. The first order was for $38 million worth of Type 56 assault rifles, 12.7mm machine guns, RPGs and lots of ammunition.

A wounded member of the Malayan Races Liberation Army (MRLA). The Malayan Campaign (1948–1960) was one of the first conflicts to involve the Kalashnikov. Prior to independence for Malaya (modern-day Malaysia), British forces fought successfully to contain Chinese-backed Communist terrorists. (*Author's Collection*)

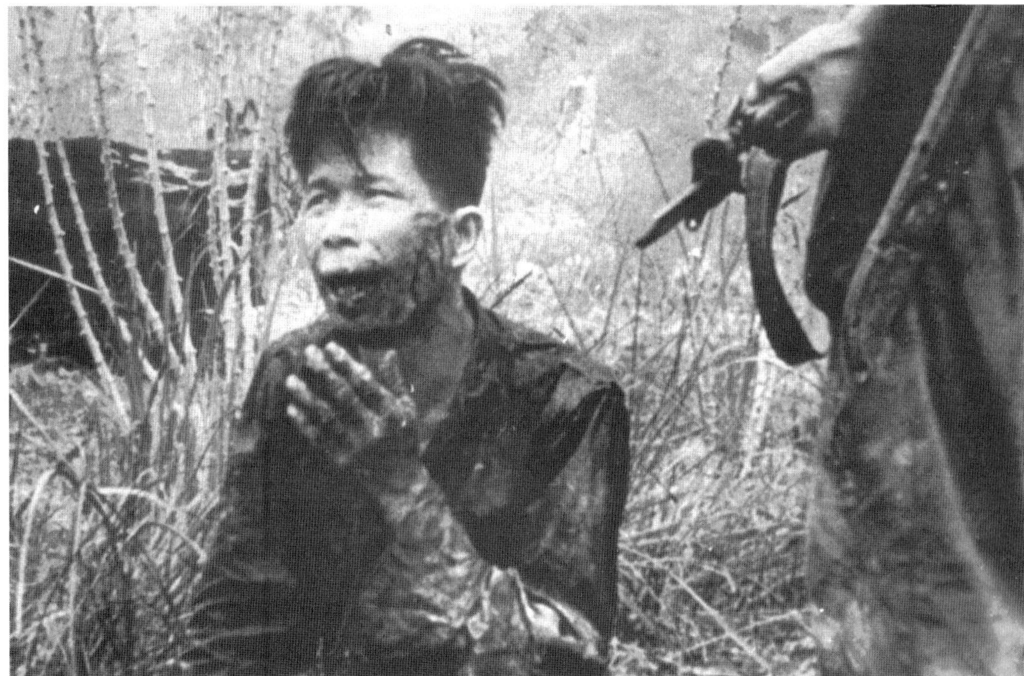

A derailed train in Malaya in the 1950s. In the decades that followed, the Kalashnikov began to arm numerous guerrilla liberation armies. The MRLA received limited numbers of the Chinese Communist or Chicom Type 56. (*Author's Collection*)

The Kalashnikov played a key role in the wars of liberation in Indo-China. Initially after the Second World War the Viet Minh and Viet Cong fighting in Vietnam had to rely on whatever weapons they could lay their hands on. The Viet Minh fighting the French relied on captured American, French, German, Japanese and Soviet guns. (*Author's Collection*)

Soldiers of the North Vietnamese Army (NVA) equipped with the Chicom Type 56 assault rifle, which was widely used by both the NVA and VC. *(Author's Collection)*

More NVA soldiers with what look to be AK-47s. Standard NVA headgear was the famous sun helmet (erroneously called a pith helmet by the Americans); the other common headgear was a floppy bush hat of the type seen here and in the previous photo. *(Author's Collection)*

NVA troops, again armed with AK-47s, on parade in their distinctive sun helmets. Chinese, Czech (Model 58 or Vz.58) and Soviet versions of the AK were all encountered in quantity by soldiers of the Army of the Republic of Vietnam and US armed forces. *(Author's Collection)*

North Vietnamese self-defence troops armed with rifles photographed on exercise in Hanoi in the summer of 1965. China claimed that it sent up to 300,000 men to help North Vietnam; Hanoi denied this number but did not deny Chinese troops fought alongside the NVA. (*Author's Collection*)

The Soviet SKS or Chicom Type 56 semi-automatic rifle (seen here) was also widely used by the NVA and later by the Viet Cong. The light machine gun on the ground is the Soviet RPD or the Chinese Type 56; in later years the Soviet RPK, the light machine-gun version of the AK, became popular. (*Author's Collection*)

American Military Police Captain Michael Harvey inspects a captured AK-47 (note the milled receiver). Such weapons would often be passed back to US intelligence to try to determine their country of origin. The muzzle break indicates it could be Soviet, but it could equally have come from a Warsaw Pact country. (*US DoD*)

A Cambodian soldier with an AK-47. The war in Vietnam spilled over into Cambodia and Laos during the 1970s. The Cambodian Forces Armées Nationales Khmere fought the NVA and the Khmer Communists. (*Author's Collection*)

NVA with assault rifles dash past abandoned American C-7A Caribou and C-47 Skytrain transport aircraft at Tan Son Nhut air base. By 1973 all American troops had been withdrawn and two years later the NVA stormed into Saigon having overwhelmed the ARVN. The Communists also took power in Cambodia and Laos. (*Author's Collection*)

Having made its mark in Vietnam, within a decade the AK-47 was doing the same in Afghanistan. Here Mujahideen photographed in Afghanistan in the 1980s are sporting a selection of AKS folding-stock Kalashnikovs. (*Erwin Franzen*)

Another Afghan guerrilla fighter, this time with an AKM. (*Julian Gearing*)

Journalist Erwin Franzen in Jaji, Paktia Province, Afghanistan, in August 1984, posing with a Chicom Type 56-1. (*Erwin Franzen*)

Guerrilla leader Ahmad Shah Massoud in the Panjsher Valley in the spring of 1985 examining a captured Soviet AK-74/BG-15 assault rifle/40mm under-barrel grenade launcher combination. Massoud was assassinated by al-Qaeda just prior to 9/11. (*Julian Gearing*)

Soviet troops take a break from the fighting in Afghanistan. The man on the left is resting his AK-74 on his knee; this became the standard Soviet small arm in the mid-1980s. The CIA paid up to $5,000 for examples of this weapon captured by the Mujahideen. (*Igor Bondarets*)

A truck-load of Afghan soldiers: the AKM was their standard weapon. (*Igor Bondarets*)

Soviet Spetnaz armed with AK-74s; the man in the middle has a PKM. The Kalashnikov design team also came up with the Pulomet Kalashnikova which served as the Soviets' medium machine gun using a 7.62 × 54mm round; this was followed by a modernised version producing the PKM. (*Igor Bondarets*)

A motley collection of Slavic and Central Asian recruits serving in the Soviet armed forces in Afghanistan. The two weapons visible are AK-74s. (*Igor Bondarets*)

The two men nearest the camera are holding the Soviet AKSU-74, also known as the AKR Krinkov. This weapon has a greatly shortened barrel and gas rod, as well as the AKS-74's metal folding stock. In order to reduce the blast, a specially large muzzle brake was added to the threaded barrel. In Afghanistan Soviet Special Forces, paratroops and armoured vehicle crews favoured this weapon. The third man has the AK-74 with the BG-15 40mm grenade launcher seen earlier. (*Igor Bondarets*)

(*Opposite, top*) Soviet troops in Afghanistan take a break from a mountain patrol. They are equipped with the folding stock AK-74. (*Igor Bondarets*)

(*Opposite, below*) An Afghan local is questioned by men armed with the AKS-74 and the PKM. (*Igor Bondarets*)

This caption should read 'Where are we?' The Soviet guard has an RPK or RPK-74 SAW or squad automatic weapon slung over his shoulder. The latter features a lighter muzzle brake than that on the AK-74, with an almost exact copy of the US M16 'birdcage' flash suppressor. (*Igor Bondarets*)

An Afghan with the Vz.58 assault rifle, Czechoslovakia's answer to the AK-47. Although similar in appearance, this weapon is not a clone but a completely different design. It remains in service with the Czech and Slovak armies. (*US DoD*)

Chapter Four

Middle East Mayhem

The AK-47 first came to prominence in the Middle East from the 1960s onwards in the hands of the Palestinian Liberation Organisation. Their struggle for a free homeland in the face of the creation of Israel captured the world's imagination. Since then every single Arab army, with the exception of the Western-aligned Gulf States forces, has used the Kalashnikov. In the deserts of the Middle East these armies learned to appreciate the fact that no matter how much dust and grit found its way into the AK-47, its firing mechanism continued to function.

When the winds began to whip up the scouring sands of the Sinai the Egyptian Army was glad that the Kalashnikov was a weapon that could be relied upon – Egypt's generals, of course, were another matter. The Iraqi and Iranian armies found the same to be true during their protracted eight-year-long border war – the dust and sands of Khuzestan and the al Faw peninsula posed few problems for the AK-47 when many other firearms regularly jammed up.

During the 1967 Six Day War the Israelis found the shorter range of their Uzi light submachine gun was not a great disadvantage against the Syrians' AK-47s. The Uzi had been first introduced with the Israeli Special Forces in the mid-1950s. The smaller and quicker-loading Uzi gave them a distinct advantage in clearing Syrian bunkers on the Golan Heights. It provided a similar edge when fighting in Jordanian defensive positions. However, during the 1973 Yom Kippur War the Uzi succumbed to the sands of the Sinai desert, unlike the Egyptians' AK-47s.

There are two key indigenous versions of the Kalashnikov in the Middle East, namely the Egyptian Misr and the Iraqi Tabuk, both of which have been widely exported. Egypt first obtained the AK-47 in the 1950s and by the end of the decade had set up Factory 54 to produce the AKM. This comes under the Maadi Military & Civil Industries Company run by the Egyptian government's Military Factories General Organisation.

There is very little to distinguish the Maadi weapon from its Soviet predecessor other than the Egyptian receiver markings. The Misr AKMS with its very distinctive single stem wire stock is a common sight in Afghanistan, while the Tabuk is still in service with the Iraqi police and to a lesser extent with the Iraqi Army.

As well the Kalashnikov and its variants, Egypt also manufactures the Soviet-designed 7.62mm, 12.7mm and 14.5mm machine guns. Large numbers of these were exported to Iraq during the Iran–Iraq War. Egypt received sizeable orders from Iraq for small arms, ammunition and armoured vehicle spare parts in the early 1980s. Weapons were also sold to Oman, Somalia, Sudan and North Yemen. The Egyptians claim that the Russians subsequently undercut their prices, a move that saw Egyptian exports drop from $1 billion in 1982 to $500 million the following year; this was also in part due to Iraq buying Western weapons. North Korean clones of the AK-47 may also have been shipped to Iran during the Iran–Iraq War. The Maadi Misr assault rifle saw extensive service during the Arab–Israeli Wars and during the 1991 and 2003 Gulf Wars. Iraq produced its own copy of the AK-47, derived from the Yugoslav Zastava M70, known as the Tabuk. This saw action during the Iran–Iraq War and in the two Gulf Wars. Iran also produced its own version, known as the KLS.

Initially the bulk of the Kalashnikovs were imported into the region from the Soviet Union and other Warsaw Pact members. At the height of the Arab–Israeli conflicts the Soviet Union supplied millions of dollars worth of arms to its Arab allies, most notably Egypt, Iraq and Syria. Small arms, of course, included the AK-47/AKM, RPD, RPK and RPG-7. These Soviet weapons were instrumental in the three Arab–Israeli Wars in 1956, 1967 and 1973, also to a lesser extent in 1982 in the Lebanon.

In both the 1967 and 1973 conflicts the Israeli Defence Force captured tens of thousands of small arms from the defeated Jordanian, Egyptian and Syrian Armies. Intriguingly, when the Israelis overran the PLO weapons dumps in southern Lebanon in 1982 the IDF found that they had not solely relied upon Soviet bloc sources. Among the ubiquitous AK-47/AKM and SKS stocks they found American M16s, which had presumably come via commercial channels.

The Israelis were so impressed by the AK-47's rugged qualities that it became the preferred weapon of the Israeli Defence Force's elite units. It also led to the development of the 5.56mm Galil ARM and SAR assault rifles. Ironically, many of the Soviet small arms captured from the Arab armies by the Israelis ended up in the hands of the Mujahideen fighting the Soviet Army in Afghanistan. The Egyptians also shipped Soviet-designed weapons to the Mujahideen.

A wide range of AK-47 assault rifles, both those produced locally and those procured from Warsaw Pact states in their many variants, became commonplace throughout the armies of the Middle East. As a consequence, the Kalashnikov found its way into the hands of the PLO, operating first out of Jordan and then Lebanon.

During the Cold War vast quantities of Warsaw Pact-manufactured weapons also poured into the Middle East. Notably after the Second World War Czechoslovakia was permitted to produce Soviet tank designs largely for export to the Middle East. The famous T-55 was licence-built from the mid-1960s for domestic as well as export purposes, followed by the T-72 in the late 1970s. Czechoslovakia also exported thousands of its OT-64 armoured personnel carriers, Iraq being the biggest customer.

Similarly Poland built Soviet tanks, while Bulgaria and Poland manufactured the Soviet MT-LB multi-purpose armoured personnel carrier, numbers of which were exported to countries such as Iraq. Romania manufactured T-55s from the late 1970s. Along with these armoured vehicles inevitably came East European-produced versions of the Kalashnikov in their thousands.

During the Iran–Iraq War North Korea was encouraged to supply Iran with T-62 tanks, even though Iraq had long been a Soviet ally. It is quite possible that the North Korean Type 58 and Type 68 assault rifles came with them. On the eve of the war the Iranian armed forces were mainly equipped with the German Heckler & Koch G3, but Ayatollah Khomeini's Revolutionary Guards were equipped with the Kalashnikov. Where they came from is unclear, though the most likely culprits were China, Libya, North Korea and Syria or the black market. Similarly many East European countries exported vast quantities of licence-built Soviet armour to the Iranians. When Iran invaded Iraq, Moscow ended its neutrality and began to supply Iraq again.

The Soviet Union's huge arms exports did not give Moscow any great strategic leverage. Egypt eventually defected back to the American camp, while Libya and Syria became dangerous liabilities; Libya moved to rehabilitate itself, but then fell into a state of civil war. Similarly Egypt, Syria and Yemen became stricken by popular unrest. Moscow's weapons did nothing but contribute to the endemic mayhem in the Middle East. The Kalashnikov was inevitably at the forefront of all this.

The Kalashnikov made its presence felt in the Middle East after it was supplied to most of the region's Arab armies and guerrilla forces, such as the Palestinian Liberation Organisation. These Palestinian fighters are armed with AK-47s and a 7.62mm Soviet Goryunov machine gun (which was also licence-built in Egypt). (*Author's Collection*)

The Kalashnikov series of weapons became the main personal small arm of most of the PLO groups; RPDs and RPG-7s were the most common support weapons. The milled receiver of the AKS-47 is clearly visible on the left. The man in the background has an SKS Simonov semi-automatic carbine. (*Author's Collection*)

This photo appears to show Egyptian troops rounding up civilians. Egypt deployed the Kalashnikov in its wars against Israel in 1967 and 1973; it also produces its own variant of the AKM known as the Misr, which has been extensively sold abroad. (*Author's Collection*)

Just prior to the war breaking out with Israel, these Egyptian soldiers are digging in on the Sinai Peninsula in the summer of 1967, having stacked their AK-47s with fixed bayonets. The Kalashnikov happily withstood the ravages of the Sinai's desert sands. (*Author's Collection*)

Another PLO fighter posing for the camera with an AKS-47. By the early 1980s the key PLO faction was Al-Fatah, which numbered some 13,000 men under Yasser Arafat. The Israelis drove them from southern Lebanon in 1982 during a six-day operation. (*Author's Collection*)

A separate Palestinian group was the Syrian-backed Palestinian Liberation Army. Never entrusted with more than support duties during the Middle East Wars, after 1973 this in effect became part of the Syrian Army and served in Lebanon. These men also appear to have AKS-47s. (*Author's Collection*)

More PLO fighters with AK-47s, this time on the streets of the Jordanian capital in mid-September 1970. This became known as 'Black September' after the Jordanian Army physically drove the PLO out, forcing it to relocate to southern Lebanon. The man in the middle is handling a Yugoslav M57 anti-tank weapon. (*Via Author*)

An Egyptian soldier firing the RPD 7.62mm light machine gun. This weapon also saw extensive service during the Arab–Israeli Wars. (*US DoD/Capt Mark Berberwyck*)

This weapon, in the hands of an Afghan soldier, is an Egyptian Misr produced by the Maadi Engineering Industries Company factory in Cairo. Apart from the factory markings, the fixed-stock Misr is completely indistinguishable from the AKM. However, the version with the folding triangular metal shoulder stop seen here is very distinctive. (*ISAF*)

Iraqi generals watch as a woman strips down a Romanian AIM. During the 1980s Iraq produced its own Kalashnikov, known as the Tabuk, based on the successful Yugoslavian M70 not the Soviet AKM. (*Author's Collection*)

Iraqi commandos training on the southern front in October 1983 during the Iraq–Iran War – their weapon of choice is the AKM. Both sides made extensive use of this weapon. (*Via Author*)

This young Iranian recruit in the Hamrein Mountains in November 1982 has a picture of Ayatollah Khomeini adorning his AKM. Iran produces a version of the Chinese Type 56, which is known as the KL-7.62. This is an unlicensed copy of the weapons China supplied during the Iran–Iraq War – more recent versions have plastic stocks and hand guards. (*Via Author*)

An Iraqi soldier manning a defensive position with an RPK in the Iranian city of Khorramshahr in October 1980. The Iranians suffered very heavy casualties before they finally drove the Iraqis out. (*Via Author*)

Another Iraqi soldier, this time with an AKM. (*Via Author*)

Iraqi troops posing for the camera. They seem to have a solitary assault rifle between them. The Iranians' costly 'human wave' tactics appalled the Iraqis. (*Via Author*)

Egyptian soldiers armed with indigenously produced 7.62mm Maadi Misr assault rifles (distinguishable in this case by their folding wire stocks) practising a beach assault at El Omayed, Egypt, during an amphibious exercise in October 2001. (*US DoD/SRA D. Myles Cullen*)

Afghan National Army soldiers in Gayan District, Afghanistan, during the first ever democratic elections held in the country in October 2004. The ANA are notable users of the Egyptian Misr, as seen here. (*US DoD/Specialist Jerry T. Combes*)

Afghan troops on patrol with US Marines and US soldiers, sporting a mixture of AKMs and Misrs. The soldier on the right seems unhappy at having his photo taken. (*US DoD/Sgt Teddy Wade*)

This recruit from the Iraqi National Police, pictured on patrol in Duwebb in February 2008, is armed with an Iraqi-produced Tabuk or a Yugoslav M70. The Tabuk is a straight copy of the M70 – thereby making it a copy of a copy! (*US DoD/Petty Officer 1st Class Sean Mulligan*)

An Afghan policeman with an Egyptian Misr that has been fitted with a grenade launcher. In Afghanistan the police are more like militia than regular policemen, and are often required to fight as infantry. (*ISAF*)

This Iraqi soldier is armed with the rather ancient Czech Vz.58 7.62mm assault rifle, which first came into service in 1958. Like the AK-47, it has seen extensive combat around the world over the years. (*US DoD*)

Chapter Five

Cold War Bonanza

By the 1950s the AK-47 had spread to Moscow's allies in the Eastern Bloc Warsaw Pact. Furthermore, many states such as Bulgaria, East Germany, Hungary, Poland and Romania had begun to produce it under licence during the 1960s. Like the Soviet Union, many of them exported vast quantities of Kalashnikovs around the world. Over the past sixty years about thirty foreign manufacturers have produced an estimated one hundred million Kalashnikovs.

Early Bulgarian-manufactured AK-47s and AKMs are identical to their Soviet counterparts and many were assembled from kits rather than actually produced in Bulgaria. The only real distinguishing features are the manufacturer's markings (a double circled 10 mark) signifying the Bulgarian State Arsenal.

The East Germans designated their versions the MPiK (Maschinenpistole Kalashnikova) and the MPiKS for the folding-stock type. The AKM became the MpiKM (M = Modern). The East Germans also produced their own side folding stock AKM known as the MPiKMS72. The wire stock is very distinctive and is very similar to the one on the Romanian version.

In 1990 Germany produced a new AKMS version, itself a variant of the MPiKMS machine pistol that was produced by East Germany under licence. At just 56cm long, this AKMS is the shortest of all the AKM variants and is essentially a cheap and cheerful answer to the AKSU-74. Likewise, Romania produced a PM md 90 carbine designed for tank crews and Special Forces at roughly the same time.

The Hungarians came up with two versions of the AKM, called the AKM-63 and the AMD-65. These have very distinctive features and are easy to identify. The AKM-63 has a plastic stock (light blue/ grey in early models and green/black in later ones), matching plastic pistol grip and a forward pistol grip (identical to the fore grip but reversed) under the fore grip, which has a characteristic ventilated metal hand guard (rather than the usual wood or plastic). The AMD-65 is a shorter version of the AKM-63 and is easily identified by a single stem side folding stock, two green plastic pistol grips, a shortened barrel and bulbous muzzle brake. Although designed for paratroopers, it is a common sight in Afghanistan and Iraq, having been exported in some quantity. During the 1970s the AMD-65 was sold to Angola and Mozambique and saw combat there during the civil wars. Some were also supplied to Zimbabwean guerrillas. It subsequently saw action during the Lebanese civil war and is still in service in Afghanistan, Croatia, Iraq and Sudan.

The Polish Lucznik Karabinek derivative of the AK-47 is known as the kbkAK. Poland produced a version of the Soviet AKM from the early 1960s and, although designated the kbkAK, it is generally better known as the PMK in the West. This can be fitted with the LoN-1 grenade launcher capable of firing the DNG-60 anti-tank grenade.

The Yugoslav arms factory of Zavodi Crvena Zastava, now in Serbia, produced Kalashnikov derivatives for the former Yugoslav People's Army and for export. The fit and finish on the Yugoslavian/Serbian guns are much better than on most other Communist-manufactured Kalashnikovs and a number of versions were produced. The early ones are known as the Model 64A/64B (fixed and folding stock respectively, based on the milled receiver). The later M70 employed both milled and stamped receivers. The much more common M70 B1 and AB2 are copies of the AKM. Zastava also produces the successful M72, which is a derivative of the RPK light machine gun.

The Czechs produced a similar-looking assault rifle called the Model 1958 or Vz.58. However, internally this weapon is quite different from the Kalashnikov and it is not a copy. The principal differences are the bolt, trigger and selector. Because of its appearance it does get confused with the AK-47, and has likewise been widely exported.

Moscow argued that after the collapse of the Soviet Union and the Warsaw Pact the Kalashnikov manufacturing licences were no longer valid. In the late 1990s the Izhmash factory in the city of Izhevsk in the Urals Mountains was granted the state patent by the Russian Federation. This was clearly a case of shutting the stable door after the horse had bolted.

Many East European countries, particularly the former Czechoslovakia, Bulgaria and Romania had warehouse after warehouse of Soviet-designed weapons that they were only too happy to shift if the price was right. While some of these had been supplied by the Soviet Union, much of it had been produced indigenously with Moscow's approval during the Cold War.

Russia's arms export company Rosoboronexport announced at the end of 2009 that it would be stepping up its struggle to protect the copyright of the Kalashnikov. To date India, Kazakhstan and Venezuela are the only legitimate licence holders. Izhmash estimates it is losing approximately $400 million a year to counterfeit Kalashnikov producers. The key culprits are Eastern European countries and China.

Many former licence holders continued to produce the Kalashnikov in one shape or another, despite the Russian Federation arguing that licences issued by the former

Soviet Union were null and void. Most notably, the Bulgarian Arsenal Company at Kazanlak manufactures the AR (7.62 x 39mm) and AR-M1 (5.45 x 39.5mm), both of which Moscow considers illegal copies. According to Russian figures, annual illegal sales of unlicensed small arms on the international market total around $2 billion, with up to 90 per cent consisting of counterfeit Kalashnikovs – many of them produced by the Bulgarians and Chinese.

Staff overseeing the United Nation's conventional weapons register in New York have long been well aware of the havoc small arms exports play with regional conflicts, but the UN is reliant on transparency and peer pressure to curtail destabilising small arms sales. While no one was looking, Eastern Europe off-loaded considerable quantities of weaponry to anyone who wanted it, thereby fuelling a post-Cold War weapons bonanza.

A Romanian soldier with an AKMS or PM md 65 (Puşcă Pistol Mitralieră model 1965), which confusingly was sold abroad as the AIMS. The earlier PM md 63 has a wooden stock and forward-facing fore grip. The later 1986 model is chambered to the 5.45 x 39mm round and has a side folding wire stock. These weapons, along with the Hungarian AMD and Bulgarian AKS, have been exported all over the world. *(US DoD/Staff Sgt Perry Heimer)*

Romanian troops on exercise with PM md 65/86s and the RPK light machine gun known as the md 64. The RPK has proved to be a highly successful and enduring squad support weapon, or in American parlance a SAW – Squad Automatic Weapon. (*Lieutenant-Colonel Dragoş Anghelache*)

Polish troops on a Warsaw Pact exercise with the indigenously produced kbkAKM and the PKM medium machine gun. All the Warsaw Pact members produced their own versions of the Kalashnikov, each with subtle distinctive characteristics. (*Graham Thompson*)

This Ukrainian marine, photographed in 1996, is brandishing the AKSU-74, which first saw combat in Afghanistan with the Soviet Army. The AKSU-74 proved far more successful than the AKM carbine, which never really caught on. (*US DoD/LCpl M.A. Sunderland*)

These Afghan soldiers are armed with a variety of Kalashnikov clones and this photo clearly shows how the many types have become mixed up. The man closest to the camera has an Egyptian folding stock Misr AKMS, the man in the middle has the Yugoslav M70/Iraqi Tabuk copy, while the last weapon is a folding-stock Romanian PM md 65. Interchangeability of parts and ammunition means that mixing such weapons does not cause problems. (*US DoD/SSgt Gary A. Witte*)

Mongolian soldiers on a United Nations training exercise near Ulaanbator, Mongolia, in the summer of 2007. This soldier's personal weapon is an RPK-74. (*US DoD/Dustin T. Schalue*)

More Romanian Kalashnikovs, this time the PM md 63 in the hands of Afghan soldiers outside Kabul. This weapon has seen combat in recent years in Afghanistan, Iraq and Libya. (*ISAF/TSgt Laura K. Smith*)

The Hungarian AMD-65 is now a little long in the tooth but is still very popular with the Afghan Police (as seen here). (*ISAF/Chief Petty Officer Julian Carroll*)

With CIA dollars backing the Mujahideen in Afghanistan in the 1980s it was not just the Chinese, Soviets and Egyptians who provided AK-47s – many Warsaw Pact members followed suit. The net result is that most AK-47 variants can be found in Afghanistan. (*Erwin Franzen*)

This Iraqi soldier has what looks to be the later Romanian model 1986, which fires the smaller 5.45mm round. (*US DoD*)

ANA troops under instruction with Romanian Kalashnikovs. (*ISAF*)

Iraqi police photographed in Baghdad in late October 2008. They have clearly been armed with ex-President Saddam Hussein's remaining weapons stocks. By the late 1980s he had almost a million men under arms – the vast majority of them equipped with AK-47s. The fore grip identifies these once more as Romanian supplied, and the lack of an ammunition magazine shows that this was a dry-firing exercise. (*US DoD*)

An Afghan policeman holding his AMD-65 in a rather strange pose. The purpose of the tape on the muzzle brake is unclear. While this variant is popular with the Afghan police, it is rarely seen in use with the Taliban or other militant groups in Afghanistan. It has reportedly been exported to Gaza and the West Bank. *(ISAF/SSgt Dayton Mitchell)*

Afghan policewomen being trained on the AMD-65. The nearest weapon has the earlier, wooden, fore grip. *(ISAF/SSgt Sarah Brown)*

Czech soldiers firing the indigenous Vz.58. Before Czechoslovakia split into two countries, this weapon was widely exported and is often mistaken for the AK-47. (*US DoD*)

This Afghan National Civil Order police officer is equipped with a Chinese Type 56-1, easily identifiable by the smooth muzzle and lack of muzzle break. This weapon was probably supplied during the Soviet–Afghan War of the 1980s, making it, like many Kalashnikovs in Afghanistan, at least a quarter of a century old. (*ISAF/Tech Sgt Mike Tateishi*)

A US Marine test-firing an RPD. This is another very common weapon in Afghanistan, having seen action during the Soviet–Afghan War, the Afghan Civil War and Operation Enduring Freedom. (*US DoD*)

This Afghan soldier is firing the SVD-137 Dragunov sniper rifle. The Soviet Army adopted this weapon in 1963; although designed by Yevgeniy Feodorovich Dragunov, it differs very little from the standard Kalashnikov. Like all post-Second World War Soviet small arms, it has been extensively copied. (*ISAF/Sgt Mathew Moeller*)

Romanian troops from the 26th Infantry marking the end of their peacekeeping duties in Iraq at Contingency Operations Base Adder in June 2009. They are armed with the Romanian version of the AK-74, which replaced the 1963 and 1965 Models. (*US DoD/SSgt Brendan Stephens*)

The Taliban have had no trouble securing weapons, as this haul of assault rifles, medium machine guns and rocket-propelled grenades testifies. (*US DoD*)

Men from the US 82nd Airborne Division examine a stash of folding-stock Kalashnikovs at an Afghan Military Force compound in August 2002 to make sure they do not have any unauthorised weapons. (*US DoD/Spc Marshall Emerson*)

A British soldier guards some of the massive stockpile of weapons that were captured from the Taliban in 2001/2002. Since then, preventing the Taliban and other factions from smuggling arms and ammunition into Afghanistan has proved an impossible task. (*UK MoD*)

Iraqi soldiers firing their RPKs during a small arms weapons instructor course taught by the US 2nd Marine Division at Habbaniyah, Iraq, in December 2005. Note the distinctive paddle-shaped stock. The Yugoslav M72, an RPK copy, utilises a regular AK stock. (*US DoD*)

Chapter Six

Guns of the Taliban

The British Army's most recent first-hand experience of the Kalashnikov is in Afghanistan. The longer range of the 7.62mm Kalashnikov compared to British and American weapons could have been a major issue but for the poor marksmanship of the Taliban fighters. Soldiers have dubbed incoming rounds 'Afghan wasps'. Making light of the very real danger they pose, the nickname reflects that, though they are noisy and there may be many of them, few actually sting.

The British SA80A2 with its 5.56mm bullet is effective to about 450 metres; in contrast the AK-47's maximum range is about 1,000 metres but it is only accurate at around 350 metres. Of the AK-47, M16 and SA80, it is claimed that the latter is the most accurate. It was not until 2010 that Britain responded to the heavier-hitting AK-47 by introducing the L129A1 Sharpshooter semi-automatic rifle firing the larger 7.62mm round with a range of around 820 metres. This was the first new combat rifle for the British Army in twenty years and it is a redesignated American LM7.

The American M16 and M4 have less recoil than the AK-47, both are lighter and the M16 is effective to at least 500 metres, giving it a more accurate reach than the Kalashnikov. The M4 carbine, like its predecessor, has proved to be an excellent weapon as long as it is maintained. As well as being better shots, American and British troops can easily carry double the amount of ammunition of the average Taliban fighter. This gives them far greater firepower before they have even called on heavier support.

In Afghanistan the Kalashnikov's greatest failing is its inaccuracy. American, British and other International Stability Assistance Force (ISAF) troops who have been on the receiving end of 'Afghan wasps' have regularly experienced this. This, though, is not due to a design fault but rather to misuse. The ageing weapons are often not maintained, ammunition quality is poor (and quite often shelf-expired from Eastern Europe's vast surplus stocks) and most users are poorly trained and unable to get the best out of the weapon.

In the case of the Taliban, technology caught up with them. Once they realised that they could not win a stand-up fight using their ageing AK-47s against the multinational ISAF forces, they began to resort to deploying ever-growing numbers of Improvised Explosive Devices or IEDs.

In the years following the fall of the Taliban in 2001, Britain's Operation Veritas turned into a rolling commitment known as Operation Herrick. Since then, under the command and control of NATO, ISAF has slowly expanded its area of control out from Kabul to extend the authority of President Hamid Karzai's government. The British public really woke up to the UK's involvement in 2006 when the 3 Para Battle Group deployed to Helmand Province to help the Provincial Reconstruction Team with security and stability. Sending just 3,000 men, of whom only a third were front-line troops, supported by six heavy lift helicopters, always seemed over-optimistic at best. This force would eventually expand threefold.

Seven years after the liberation of Kabul there were thought to be around 10,000 Taliban, of whom around 3,000 were full-time fighters. The Taliban remain well armed with copius quantities of AK-47s and RPGs, and the suspicion is that fresh supplies of weapons and ammunition were being smuggled via Pakistan and China.

In Helmand and elsewhere bravery and tenacity, coupled with superior NATO firepower, soon persuaded the Taliban that they could not win a conventional battle. What is notable is that the Taliban, just like the Mujahideen before them, are never short of guns and ammunition. When they swept into power they did not lack armaments and when they began launching their escalating counterattacks on ISAF they were certainly well equipped.

Somebody somewhere in Central Asia has ensured that the Taliban are never short of AK-47s, 7.62mm ammunition, RPG-7 launchers and grenades. Their supplies seem limitless. The finger can only point at China, Pakistan and the former Soviet Central Asian states such as Kazakhstan. A few select factory owners and middlemen have become very wealthy through the suffering of others.

(Opposite, below) Two more Northern Alliance fighters: one has the 'newer' AKM while his comrade has an RPG-7. Guns and ammunition were flown in to the Northern Alliance prior to their push on Kabul. Interestingly, the AKM has the grooved stock normally associated with the AK-74. Mix and match is a common practice among the Afghans. (*Via Author*)

(Above)
A Northern Alliance fighter with his trusty AK-47. Many of the guns in circulation at the time of Operation Enduring Freedom in October 2001 had survived the Soviet-Afghan War and the Afghan Civil War which saw the Taliban come to power. (*Via Author*)

In the years since the Taliban were driven from power, NATO has struggled to get the Afghan National Army back on its feet. To this day their weapon of choice remains the AK-47 – these men are equipped with the Romanian version. The British Army takes a dim view of the fighting abilities of the ANA and ANP. (*ISAF*)

An Afghan policeman with the Romanian version of the AKM, the PM md 63. This weapon was also used extensively by Saddam Hussein's armed forces and more recently by fighters of the National Transitional Council in Libya. (*ISAF*)

An Afghan army patrol, again equipped with Romanian Kalashnikovs. The fore grip greatly helps to stabilise the weapon, especially in automatic mode; both the British SA80 and the American M4 now include similar fore grips. (*ISAF*)

Afghan troops engaging enemy targets in late 2009. The soldier in the foreground has an AK-47, while behind him an RPK gunner rests his weapon on its bipod. (*ISAF/Sgt Wayne Gray*)

An American soldier examines a Chinese Type 56 captured during Operation Mountain Sweep in August 2002. The weapons stacked to his left are even older lever-and-bolt-action rifles. Albania produced a straight copy of the Type 56 known as the Automaitku Shqiptar tipi 1982 (ASH-82), which saw action with the Kosovo Liberation Army. (*US DoD/Spc Marshall Emerson*)

Afghan National Police recruits being taught how to carry their AMD-65s while walking. The guns have wooden pistol grips rather than the normal plastic ones. (*ISAF*)

(*Opposite, top*) This photo provides a good view of the shoulder rest on the AMD-65; although a no-frills weapon, it has stood the test of time just like all the other Kalashnikov derivatives. (*ISAF*)

(*Opposite, below*) Members of Afghanistan's Counter-Narcotics Police department with a selection of rather tatty looking AKMs and AMD-65s; note the pocket lights taped on to the first two weapons. Afghan security forces and the Taliban are notoriously shy of fighting in the dark. (*ISAF*)

These ANA soldiers have AKMs slung over their shoulders; field repairs include tape on the pistol grips and the fore grips. The Americans would probably like to phase the Kalashnikov out of service in Afghanistan, as they have done in Iraq. (*ISAF/Sgt Wayne Gray*)

Another shot of the same men under instruction. The soldier on the left has an RPK light machine gun slung under his arm, identifiable by the lengthened barrel and the folded-up bipod. This remains the standard Afghan squad automatic weapon in Afghanistan. (*ISAF/Sgt Wayne Gray*)

ANA or ANP training with the Type 56 and AMD-65 at the Camp Wright firing range in Asadabad in December 2009. The course, overseen by US Military Police, was intended to increase weapon proficiency and safety! Accuracy on both sides remains a problem. (*ISAF/TSgt Brian Boisvert*)

Yet more Afghan police with the ubiquitous AMD-65. (*ISAF*)

Afghan soldiers undergoing marksman practice with AKMS and AIMS Kalashnikovs. *(ISAF/Sgt Gary A. Witte)*

Afghan police on patrol in Arghandab District, Kandahar, equipped with solid stock AKMs. The second man has the GP-25 40mm grenade launcher attachment. *(ISAF)*

A pick-up truck full of Afghan soldiers ready for action. They seem to have a mixture of AK-47s and AKMs. (*ISAF*)

A member of the 1st Battalion, 2nd Brigade, 209th Corps, Afghan National Army. He is holding an AKMS, the folding-stock version of the AKM. The ribs are clearly visible on the receiver cover. (*ISAF*)

Afghan National Police armed with the Egyptian Misr. Egypt supplied this weapon to the Mujahideen during the Soviet–Afghan War. (*ISAF*)

(*Opposite, top*) An Afghan soldier poses with his AKM – it is clearly a very old weapon, judging by how much the burnishing has worn off the receiver cover. Note the spare magazines stuffed into his chest pouches. (*ISAF*)

(*Opposite, below*) British troops with an array of captured Taliban weaponry that includes heavy machine guns, anti-aircraft guns, mortars, recoilless guns, RPGs and AK-47s. (*UK MoD*)

An Afghan soldier crawling on the ranges. Poor marksmanship has always been a problem with most Kalashnikov users. *(ISAF)*

These captured Taliban Kalashnikovs consist of two AKMs (top and bottom) and an even more ancient AKS-47 (centre). These weapons are worn and poorly maintained. The plastic cutaway magazine is designed to give the shooter some idea of how much ammunition he has left – but all it does is act as a dust and dirt trap. Also note the home-made magazine pouches incongruously decorated with flowers. *(ISAF)*

Chapter Seven

Saddam's Ghost

While the AK-47 will doubtless be around for decades to come, Iraq proved to be something of a watershed in the weapon's history. It was in Iraq that the Kalashnikov's designer received a significant snub, first from Bulgaria and then from America.

Shortly after the overthrow of Saddam Hussein there was an urgent need to arm the 40,000-strong Iraqi National Guard. Ironically, Iraq was awash with weapons but many were in a poor condition and thousands had simply vanished in the aftermath of the invasion. As the Iraqi forces were already familiar with the AK-47, it seemed sensible to re-equip the new post-Saddam forces with new ones. It seemed that Izhmash would step into the breach and supply the required AKMs, or even the AK-74M or AK-103.

The Russians, already conscious that they were losing billions of dollars a year due to the illegal production of Kalashnikov clones, were dismayed when the Bulgarians undercut them on the price. Bulgaria's Arsenal Company in Kazanlak secured an American contract in 2004 to manufacture its AR assault rifle for less than $100 a piece. The AR range of weapons comes in both 7.62mm x 39mm and 5.56mm x 45mm variants that are all characterised by milled receivers with black plastic furniture. The Russians consider the Bulgarian AR-M1 a counterfeit. Nonetheless, it was not long before the Iraqi Army was sporting brand-new Bulgarian-made Kalashnikovs on the streets of Baghdad and elsewhere.

Three years later the Americans got their revenge on Bulgaria and Russia when the AK-47 suffered the indignity of being retired from service altogether in Iraq. On the grounds that the American M16 is superior and more durable than the AK-47, Washington moved to re-equip the entire Iraqi Army with 165,000 M16A2 assault rifles and M4 carbines. It was also claimed that the different make and calibre AK-47s in Iraqi service could cause maintenance and reliability problems.

At the time of Operation Iraqi Freedom in 2003 it was estimated that there were fifteen to twenty million weapons in circulation in Iraq, many of them Soviet-designed AK-47 assault rifles. Yet remarkably the following year the newly created Iraqi security forces struggled to arm and re-equip themselves. While many of Iraq's old guns were destroyed following the invasion or were simply unserviceable, vast numbers fell easily into the hands of the growing insurgency. Iraq procured in the region of a million rifles and pistols after 2003. One may wonder why they needed such vast quantities of guns.

America knew where it could lay its hands on a very large stockpile of surplus weapons: the Balkans, despite UN efforts to get the region's governments to destroy them. Balkan assault rifles were costing US and European buyers as little as $50–100 per gun (as opposed to around $250 for brand-new ones). The consensus in Bosnia and Serbia is that the quality of the Yugoslavian and subsequent Serbian Zastava M70 assault rifle is superior to any other Kalashnikov, including those of Russian manufacture (they are better in automatic fire and the barrel heats up less).

The British and American governments quickly found themselves in the middle of a row following allegations that the weapons supply chain for the Iraqi military had been systematically abused. A network of weapons brokers encompassing America, Croatia, Germany, Moldova, Serbia, Switzerland and the UK, while profiting from lucrative contracts, quietly spirited away a quarter of a million guns. There can be no denying that Iraq turned into a weapons black hole following the scramble to rearm the Iraqi armed forces.

Also in early 2007 some 20,000 Chinese Type 56 assault rifles transited the UK en route to Iraq, despite a European Union arms embargo against China. A question was raised over the issue of trans-shipment licences, but the UK interprets the embargo as only against exports to China (Iraq subsequently signed a $100 million deal with China for small arms).

The rehabilitation of the Iraqi Army was a slow and often painful process, tainted as it was by association with Saddam Hussein's appalling catalogue of human rights abuses. Nonetheless, like the phoenix rising from the ashes, a new Iraqi Army eventually emerged to help fight the War on Terror, only this time it was armed with the M16.

In a single stroke Washington sought to expunge the Kalashnikov's Vietnam War reputation and finally end its Cold War influence in the Arab world. The symbol of revolutionary zeal and developing world ubiquity was considered redundant. There was worse to come, though, when Moscow announced that it too had decided the Kalashnikov had finally come to the end of its useful service life in 2011.

British troops examining a haul of AKM assault rifles and RPK light machine guns following the invasion of southern Iraq in March 2003 by the Royal Marines in support of Operation Iraqi Freedom. These weapons look to be in a poor condition and the men's faces say it all. They could easily date from the First Gulf War over a decade earlier. (*UK MoD*)

Like many Soviet-equipped armies, Saddam Hussein's had billions of dollars' worth of military equipment stored away in bunkers and warehouses that were then often abandoned to the elements. In the middle of the frame are the remains of what was once a Tabuk assault rifle, surrounded by anti-personnel bombs. (*Author's Collection*)

American soldiers holding up yet more Iraqi war detritus. These assault rifles, despite being in crates, have clearly been left to the mercy of the elements, with predictable results. (*Author's Collection*)

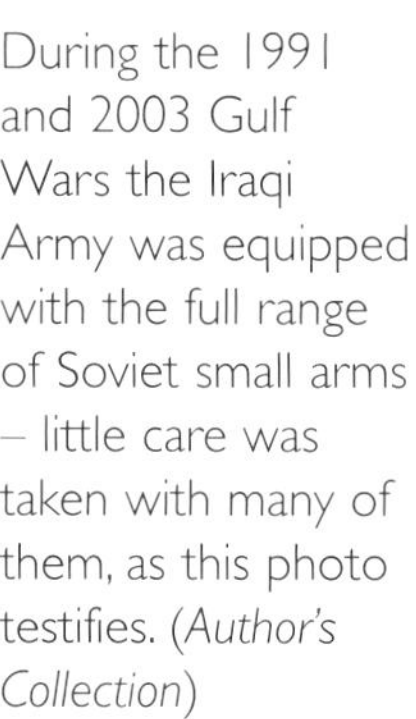

During the 1991 and 2003 Gulf Wars the Iraqi Army was equipped with the full range of Soviet small arms – little care was taken with many of them, as this photo testifies. (*Author's Collection*)

Kalashnikovs lie strewn at the roadside after a possible insurgent's pick-up truck was searched. AK assault rifles, both locally produced and obtained from the Warsaw Pact in their many variants, were the predominant infantry weapon of the Iraqi Army. These included the Romanian AKM, Yugoslav M70, the indigenous Tabuk and even possibly the North Korean Type 68. The soldier is holding the Soviet-designed PKM medium machine gun. (*US DoD*)

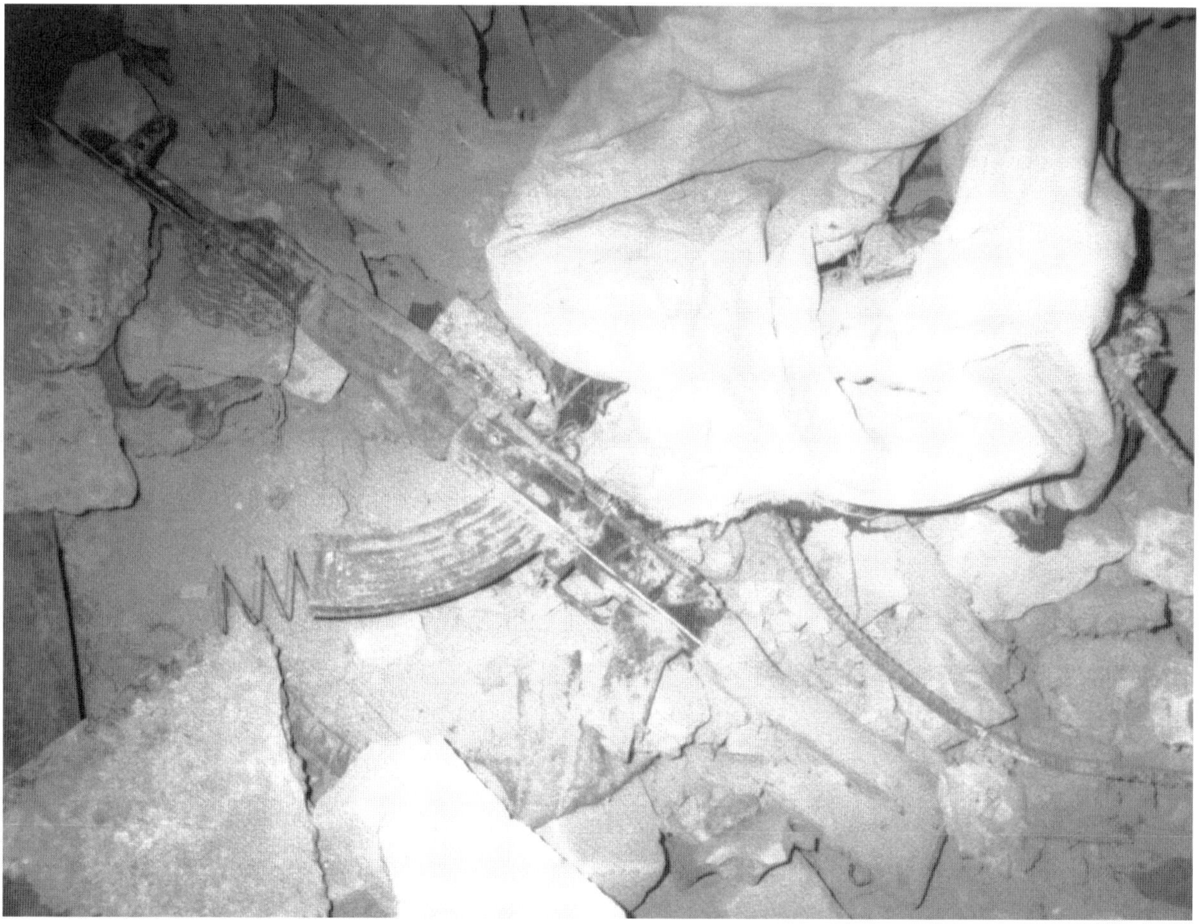

Another Iraqi AKM lies discarded among the rubble of one of Saddam's numerous weapons bunkers in 2003. (*Author's Collection*)

This blurred shot shows a pile of assault rifles and RPGs abandoned in a pond – judging by their condition, they had been in the water for some time. Iraq had little concept of firearms security. (*Author's Collection*)

Another pile of rusting and abandoned AKMs and RPGs. The Iraqi armed forces had such vast quantities of weaponry that such waste was not considered a problem. (*Author's Collection*)

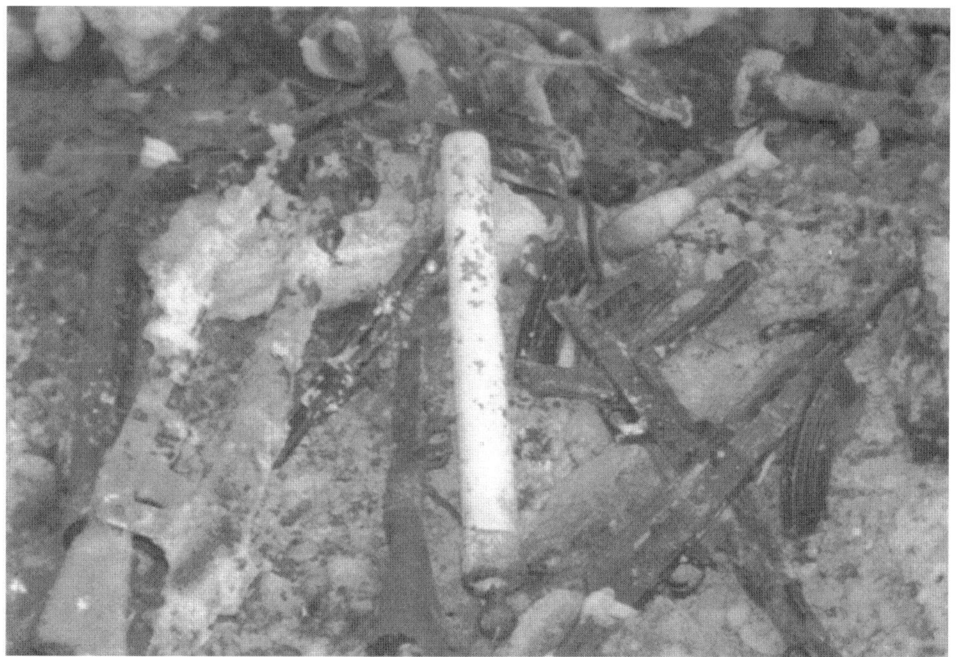

Here Iraqi Tabuks lie discarded with mortar bombs. The raised grenade launcher sight fixed to the gas tube above the barrel is visible on the weapon on the right. These were only fitted on Tabuks and M70s. (*Author's Collection*)

A mound of rusting barrels: the roof had gone, the rain got in. To the Iraqi quartermaster, this was just a pile of inconvenient rubbish. (*Author's Collection*)

Two crates of rusting Iraqi rocket-propelled grenades and AKM magazines. Normal storage practice is to grease everything then wrap it in greaseproof paper; under such conditions weapons and munitions will last indefinitely. The Iraqis cared little for such niceties. (*Author's Collection*)

Iraqi Police from the 2nd Battalion, 1st Brigade, 1st National Police Division, on patrol in the Meshra al Bawi district of Baghdad in search of insurgents and weapons caches in March 2007. The leading officer has a brand-new folding-stock Serbian M70, while the others have older-looking AKMS. The policeman in the middle, despite the machine-gun bandolier, is carrying a Dragunov sniper rifle. (*US DoD/SSgt Bronco Suzuki*)

In an effort to standardise the plethora of different types of Kalashnikov in use by the Iraqi National Guard, America ordered 40,000 Bulgarian AR-M1s to re-equip them. These Iraqi soldiers are armed with AK-47s with black plastic stocks, which are probably of Bulgarian origin. (*US DoD/Petty Officer 1st Class Sean Mulligan*)

An Iraqi soldier indulging in some 'hearts and minds' tactics in eastern Baghdad in December 2008. His weapon is almost certainly a Bosnian- or Serbian-supplied M70 manufactured by Zastava. This weapon had proved itself during the wars in the Balkans during the 1990s. (*US DoD/SSgt James Selesnick*)

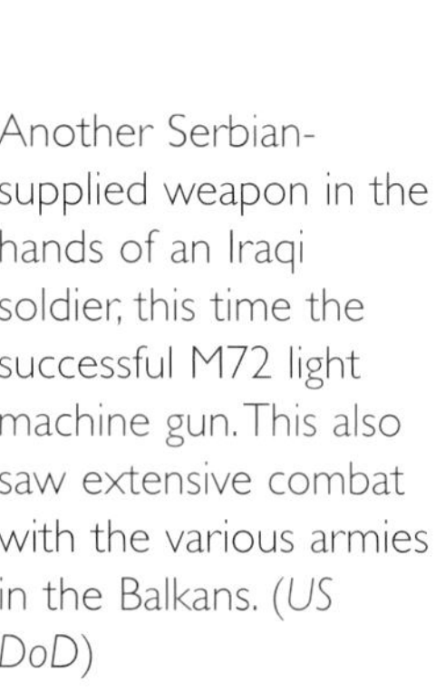

Another Serbian-supplied weapon in the hands of an Iraqi soldier, this time the successful M72 light machine gun. This also saw extensive combat with the various armies in the Balkans. (*US DoD*)

This Iraqi guard has been issued with the Czech Vz.58, which fires the same size round as the AKM, so ammunition is not a problem. As well as seeing action in the Middle East, this weapon also served extensively in Africa. (*US DoD*)

This Czech soldier, again fielding the Vz.58, is serving with the Multinational Division in Iraq. (*US DoD*)

Iraqi police celebrate after a parade in recognition of Baqouba Sovereignty Day in Diyala Province in June 2009. The gun on the left appears to be a Chinese Type 56. (*US DoD/SSgt Ali E. Flisek*)

An Iraqi policeman with an Egyptian Misr. (*US DoD/Sgt Tierney Nowland*)

Iraqi police undergoing firing training with elderly AKMs. (*US DoD*)

The Iraqi soldier nearest the camera has an M70, though it could equally be a Tabuk. The latter was manufactured in Iraq during the 1980s and 1990s, though in reality many were probably simply assembled from Yugoslav-supplied kits. (*US DoD*)

These captured Kalashnikov assault rifles and light machine guns, taken from Iraqi insurgents, clearly illustrate the vast array of different types of AK-47 that were in circulation in Iraq. (*US DoD*)

The two Iraqi soldiers on the right have fairly new AKMs. From 2004 onwards Washington purchased hundreds of thousands of new and second-hand assault rifles with which to re-equip the Iraqi armed forces. Many disappeared on to the black market. (*US DoD*)

Chapter Eight

Operation Harvest

Before the collapse of Yugoslavia, the Yugoslav People's Army or Jugoslovenska Narodna Armija (JNA) was equipped with the Zastava M70 assault rifle, the M72 light machine gun and the Yugoslav-produced SKS carbine. During the subsequent wars in former Yugoslavia between 1991 and 2001 the Bosnian, Croat, Slovene, Macedonian and Serbian forces were predominantly armed with the M70 and M72. Croat forces also had stocks of the Hungarian- and Romanian-produced AKM and initially even employed the Second World War vintage Soviet PPSh-41 and the German MP40 submachine guns.

When the fighting finally came to an end with the Dayton Accords, Operation Harvest, under the auspices of NATO, sought to encourage civilians to turn in their illegal weapons. This process was conducted by SFOR (the Stabilisation Force in Bosnia and Herzegovina) at the steel works at Zenica in Bosnia where surplus stocks were consigned to the furnace.

The process commenced in December 2002 when 3 tons of illegal weapons, including small arms, machine guns, anti-tank weapons, mortars and handguns, were melted down. The following March 3,000 small arms including M70s and M72s were destroyed at Zenica. Between February and March 2004 over 100,000 rounds of small arms ammunition and over 2,500 hand grenades were also rounded up.

However, not all of Bosnia's guns were consigned to the flames. During 2004 and 2005 the US government arranged for some 200,000 Bosnian assault rifles (as well as forty to fifty million rounds of ammunition) to be sent to Iraq, but there is no evidence that they ever arrived. Bosnian documentation for five shipments declared Coalition forces in Iraq to be the end users. A company by the name of Aerocom (registered in Moldova) airlifted 99 tons of mainly Kalashnikov AK-47 assault rifles in four flights in the summer of 2004; where they ended up is anyone's guess.

This sorry tale of woe gets worse thanks to the activities of the US DoD. According to EUFOR (the EU-led peace-keeping force in Bosnia and Herzegovina) and the Organisation for Security and Cooperation in Europe, between 2004 and 2005 contractors working on the US DoD's behalf in fact shipped over 350,000 Kalashnikov assault rifles not only out of Bosnia but also out of Montenegro and Serbia. NATO and EUFOR reportedly authorised these transfers through Iraqi intermediaries. Some guns were also sent to Afghanistan. In addition, outside former Yugoslavia a network of arms suppliers shipped guns to Afghanistan and Iraq from Albania, Bulgaria, Estonia, Slovakia and Slovenia.

The audit trail for these Balkan weapons is non-existent, with none of the former Yugoslav states admitting to the annual United Nation's register of conventional arms of selling large quantities of assault rifles. The register monitors exports and imports. During this period Bosnia and Herzegovina declared no weapons exports except for some ancient tanks and armoured fighting vehicles sent to the UK for museum purposes. In 2007 it acknowledged sending just 4,900 M70 assault rifles to Afghanistan. Likewise Serbia professes to having sold nothing apart from some artillery to Myanmar (Burma) in 2005.

In fact the Serbians claim to have been in the business of destroying weapons, not exporting them. Reportedly 100,000 arms and two million rounds of ammunition were destroyed between 2001 and 2004. However, according to the Serbian Ministry of Defence, once small arms, light weapons and ammunition become surplus they are treated as commercial goods, unless their condition dictates that they be marked for destruction. This seems like a convenient ruse for not declaring any government-approved exports. Serbia also claimed that no weapons were stolen from depots during 2004–2005. Nonetheless, in 2008 the Serbian weapons industry generated 4.5 per cent of the country's total exports, worth around $10.5 billion.

This was not the end of Bosnia's involvement in dodgy deals in Europe, for in 2006 Amnesty International discovered that 63,800 Kalashnikov assault rifles and over twenty-three million rounds of ammunition were approved for transfer to Iraq by a Swiss firm. The company denied all knowledge of such a transaction. Iraq then complicated matters further by signing a perfectly legitimate $236 million contract with Serbia for weapons, which included M21 and M70 assault rifles.

It came to light in the summer of 2007 that a British company had been permitted to import at least 40,000 M70 assault rifles into the UK from former Yugoslavia. These imports were reportedly part of the massive procurement programme designed to re-equip the new Iraq Army. This, though, was just the tip of the iceberg, as the British Parliament's Quadripartite Committee, which monitors arms exports, has reason to believe that as many as 200,000 weapons may have come into the UK without proper oversight.

Clearly the UK became a major middleman in the international small arms trade. At the same time the US government has admitted that 190,000 weapons supposedly delivered to the Iraqi armed forces and police remain unaccounted for. The upshot of all this is that an awful lot of Kalashnikov assault rifles, that arguably Iraq did not need in the first place, disappeared on to the black market and potentially into the hands of insurgents, militants, terrorists and organised crime.

An Iraqi police officer training at Forward Operating Base Hammer, outside eastern Baghdad, with a Yugoslav or Yugo M70. (*US DoD/Sgt 1st Class Alex Licea*)

This Zastava M72 or 5.56mm Tabuk light machine gun is being used by an Iraqi soldier to engage insurgents during the Second Battle of Fallujah in November 2011. (*US DoD/LCpl J.A. Chaverri*)

A Bosnian soldier posing with a rather ancient Yugoslav M56 submachine gun. The key personal weapons of the Bosnian, Croatian and Serbian armies during the Balkan Wars were the Zastava M70 and M72, but they also made use of much older weaponry dating back to the Second World War. (*Author's Collection*)

By the time the wars had ended in the Balkans the region was awash with weapons. The guns in this seized crate are Czech Vz.58 7.62mm assault rifles, which are outwardly similar to the AK-47 but are in fact a completely different weapon. It has been widely exported and although old remains in service. (*US DoD*)

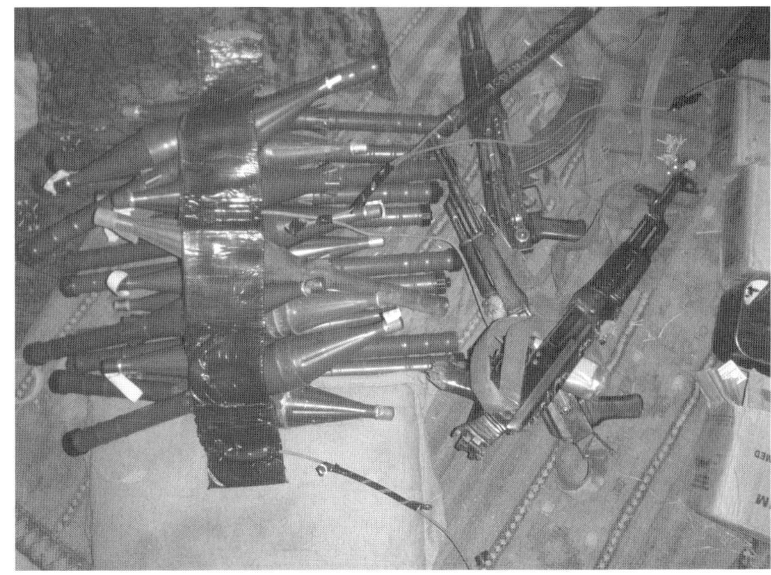

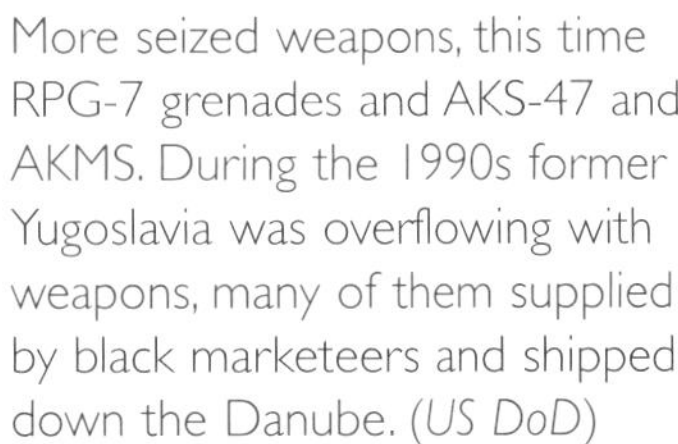

More seized weapons, this time RPG-7 grenades and AKS-47 and AKMS. During the 1990s former Yugoslavia was overflowing with weapons, many of them supplied by black marketeers and shipped down the Danube. (*US DoD*)

The furnace at the Zenica steel works, Bosnia. Thousands of illegal guns were destroyed here under Operation Harvest, but thousands more ended up on the international arms market. (*SFOR*)

Surrendered assault rifles and carbines, including AKMs, M70s and Type 56s, as well as SKS, waiting to be melted down into scrap metal. (*IFOR*)

Another pile of M70s and Type 56s. Many of the latter came via Albania. The graffiti 'War for a peace' sums up the tragic disintegration of former Yugoslavia that led to wars in Slovenia, Croatia, Bosnia and Herzegovina and Kosovo. (*SFOR*)

Russian and American peacekeepers searching for Kosovo Liberation Army weapons in Kosovo after the Serbian Army withdrew. In the late 1990s the KLA quickly sided with NATO and was disarmed under the auspices of the Kosovo Peace Implementation Force. (*US DoD*)

This Russian soldier is manning a checkpoint in eastern Kosovo in late October 2001. He is equipped with the AK-74M, which was accepted as the standard service rifle of the Russian Federation following the collapse of the Soviet Union in 1991. (*US DoD/Spc Jessie Gray*)

The Vz.58 V model with side folding stock in use with the Czech Army serving with the KFOR operation in Kosovo. (*Czech MoD*)

An Albanian soldier on guard duty in Tirana armed with a milled receiver AK-47, which is possibly of Bulgarian origin. China also supplied the Albanian armed forces with the Type 56, which Albania copied to produce the ASH-82. (*Dave Proffer*)

Another view of the same soldier, giving a clearer view of his AK-47 magazine chest pouches. The Albanian Army is only about 15,000-strong. (*Dave Proffer*)

This Iraqi policeman has been issued with an M70, which has come from either Bosnian or Serbian Army stocks. Scandalously, many of the weapons supplied by Bosnia and Serbia disappeared on to the black market before they even reached Iraq. (*US DoD/Master Sgt Jonathan Doti*)

Likewise this Iraqi Special Weapons and Tactics team member from Basra has a folding-stock M70. Some argue that this particular assault rifle is superior to the original AKM. (*US DoD*)

This haul of assault rifles, light machine guns and rocket-propelled grenades is typical of the weapons taken from both Iraqi insurgents and the Afghan Taliban. (*US DoD*)

Serbian Gendarmerie photographed in June 2009. Their personal weapons are the Zastava M21 assault rifle (the successor to the M70). This newer weapon has been exported but in nowhere near the numbers of the older M70. (*Boris Dimitrov*)

Chapter Nine

Allah and the Kalashnikov

Militant Islam is characterised by Allah and the ubiquitous Kalashnikov assault rifle. Indeed, Islamic militants around the world have never experienced real difficulties in obtaining weapons thanks in part to the arms markets at Bakara in Somalia, Peshawar and Quetta in Pakistan, and Souq al-Talh and Ma'rib in Yemen. Certainly Pakistan never had a problem arming the Mujahideen and the Taliban.

High in the snowy mountains of the Hindu Kush and down in the dusty valleys of Afghanistan, the Mujahideen learned the same lesson the Viet Cong had in Vietnam: that regardless of the climatic conditions, the Kalashnikov just kept on firing. The Soviet Army was none too happy at being on the receiving end of its own invention. When the Taliban swarmed across the Pakistani border, their preferred weapon was once again the AK-47.

By the mid 1980s some 400,000 CIA-funded weapons had been supplied to the Mujahideen (albeit over a quarter of those provided by Egypt and India were either obsolete or in poor condition). Former Palestinian Liberation Organisation weapons captured by the Israelis in Lebanon were also supplied via the CIA (Israel's haul had included almost 27,000 guns). Turkey provided the Pakistani intelligence service, the ISI, with 78,000 small arms but the bulk of these were so faulty they were not handed over to the resistance. In addition, 65,000 tons of ammunition was passing through Pakistan a year. It is conceivable that by the time the Soviets withdrew somewhere in the region of two million weapons had gone into Pakistan and Afghanistan. Many of these are still in service today.

Thanks to the Mujahideen, American and British intelligence got their hands on the then relatively new AK-74 and the RPK-74 during the 1980s. The latter had been found to be inferior to the American M249 Squad Automatic Weapon. Fired from the prone position, the RPK-74 was found to high-centre on the 45 round magazine and lift off the bipod, making the weapon unstable. There was also some amusement that the muzzle brake was a copy of the 'bird cage' flash suppressor found on the American M16. Likewise, the trigger and firing mechanisms were found to have been copied directly from the American M1 Garand rifle.

The AK-74 and RPK-74 were the logical progression of the Kalashnikov family and are essentially AKMs chambered for the then new 5.45 x 39mm Communist Bloc or ComBloc cartridge. The RPK-74 magazine is identical in construction to the early AK-74 and later AKM magazines. Made in two parts from glass-reinforced, rust-coloured polyethylene plastic, it is glued together with a viscous, two part epoxy-resin adhesive. With a protective metal plate on the base, these magazines have great strength and durability and last for years. In the case of the RPK-74, it can also take the thirty-round AK-74 magazine and a hundred-round plastic drum magazine.

The illegal arms trade run from the weapons markets at Souq al-Talh and Ma'rib, both of which were reportedly shut down, have fuelled terrorism in Yemen. The country is also recognised as the principal arms supplier to the Somali warlords and pirates, and is the main source of weapons for the Bakara arms bazaar. Much of this weaponry came via Afghanistan and Pakistan. Notably, the AK-47 and the RPG-7 are the chosen weapons of the Somali pirates who terrorise the waters off the Horn of Africa.

In recent years Egypt became a major smuggling conduit in the Middle East for large quantities of illegal Kalashnikovs and other weapons being delivered to Palestinian militants in the Gaza Strip and West Bank via tunnels beneath the Gaza–Egyptian border. At its peak 3,000 rifles and 200 RPGs were being brought over every month. During 2005 200 anti-tank rocket launchers, 350 rockets, 5,000 automatic rifles and over a million rounds of ammunition were smuggled into Gaza. By the end of 2006 these figures had risen to 20,000 assault rifles, 3,000 pistols and six million rounds of ammunition, according to Israeli intelligence sources.

The Egyptian authorities claimed the bulk of the arms were coming from Israel, intent on destabilising the Palestinians. Ironically, in 2006 Israel found itself providing 3,000 rifles and three million rounds of ammunition to its old enemy Fatah (the former guerrilla wing of the PLO) to support their confrontation with Hamas, the ruling party in the Gaza Strip.

A Palestinian Liberation Organisation or Palestinian Liberation Army fighter with a milled-receiver AK-47. PLO training regularly featured live-firing exercises that inevitably resulted in accidents. Over the decades the Israeli Defence Forces have captured tens of thousands of these weapons from the Egyptians, Syrians and Palestinians; many were reused or sold on. (*Author's Collection*)

This slightly damaged photo shows PLO Fatah fighters posing for the movie cameras; again they are armed with the older Kalashnikov, in this case folding-stock AKS-47s. The Israelis liked the weapon so much that their Special Forces adopted it. (*Author's Collection*)

Two Mujahideen in the 1980s. The young man on the left has an old Enfield rifle, while the older man has the Chicom Type 56-1. China shipped weapons via Pakistan, but was also able to smuggle them across the Taklimakan Desert and down the Wakhan Corridor, which links China directly with Afghanistan. *(Julian Gearing)*

Weapons captured from the Mujahideen. The two tubes are disposable light anti-tank weapons, while the Kalashnikov is clearly the Chinese Type 56. The presence of this weapon made it difficult for China to deny its involvement in the Soviet–Afghan War. *(Author's Collection)*

Following the Soviet–Afghan War, Algeria was one of the first places to experience the fall-out when Algerian Jihadists returned home to try to create an Islamic state there during the 1990s. These Algerian police are about to conduct a raid on a militant hide-out. The officer's firearm is an AKS-47 or AKMS. (*Via Author*)

Likewise, the Chechen separatists seen here were supported by Islamic militants, who had experienced combat fighting against the Soviets. The Chechens fought a series of particularly brutal wars against the Soviet Army. (*Via Author*)

These Russian security forces fighting in Chechnya are armed with a Dragunov sniper rifle (*left*) and an AK-74M. Chechen forces regularly used the same weaponry. (*Via Author*)

Kashmiri separatists trained in Pakistan and Afghanistan were extensively armed with the Kalashnikov. (*Via Author*)

This Pakistani soldier stands guard over an enormous cache of weapons taken from Indian-backed Kashmiri guerrillas. On the ground are PPSh-41 submachine guns (probably the Chinese copy known as the Type 50); behind them are stacked Dragunov sniper rifles and a huge array of mostly folding-stock Kalashnikovs. Note the 75-round drum magazine fitted on two of them. (*Via Author*)

In recent years the Pakistani Army has had to wage a series of campaigns against the Pakistani Taliban operating in the provinces bordering Afghanistan. Most of these soldiers have self-loading rifles, but the NCO has an AKMS. (*Via Author*)

This Afghan Northern Alliance fighter is clasping an AKM. The two men on the right have an AKM and AKMS respectively. (*Via Author*)

An AKMS and explosives seized from the Taliban by the Afghan National Police. IEDs soon replaced the AK-47 and RPG as the Taliban's weapon of choice in the war against ISAF. (*US DoD*)

Yet more weaponry seized from the Taliban. The haul includes assault rifles, medium machine guns and RPGs, which were used to launch attacks on US forces in Afghanistan. (*US DoD*)

Somewhat unusually, this Afghan policeman engaging Taliban targets is making an effort to aim his weapon, rather than employing the preferred 'spray and pray' technique. The weapon is an old Type 56-1 with wooden furniture, and the folding stock has been decorated with tape. (*ISAF*)

Accuracy has never been a hallmark of the Taliban or the Afghan security forces. These Afghan police are undergoing battle drills at Forward Operating Base Ramrod, Afghanistan, in late February 2010. From left to right they are equipped with an RPK light machine gun, a Misr and AMD-65 assault rifles respectively. (*US DoD/SSgt Dayton Mitchell*)

The Taliban are never short of weapons: on the right are PKM medium machine guns, plus two RPGs and an assault rifle. Grenades for the launchers are always available in abundance. Stopping guns and ammunition getting into Afghanistan is an impossible task. (*US DoD*)

Chapter Ten

The AK-100 and Beyond

Ultimately, it was economic not military assistance that many of Moscow's weapons clients really needed. This was something Russia was unable to provide and ironically the gifting of billions of dollars' worth of arms, including the AK-47, contributed to the collapse of the Soviet Union and bankrupted its arms industry. Moscow's Kalashnikov diplomacy ended in complete failure. While there can be no denying that the Kalashnikov became an icon of the many regional conflicts fought during the Cold War, it was at a terrible price for all concerned.

The announcement in September 2011 that Moscow would not be ordering any more AK-74M assault rifles must have come as a shock to its designer, Mikhail Kalashnikov. He had long been fêted as a national hero, and variants of his ubiquitous AK-47 have been in service around the world for the past sixty-four years. However, the time had finally come for a brand-new standard Russian Army assault rifle.

In fact, the Kalashnikov had been on borrowed time in Russia for some years. Project Abakan saw new assault rifle designs being trialled in the Soviet Union throughout the 1980s. In these trials designer Gennadiy Nikonov's Avtomat Nikonova Model 1994 (AN-94) out-performed all its rivals. Chambered for the same 5.45mm x 39mm cartridge as the AK-74, this weapon was declared the successor to the Kalashnikov family of weapons in the mid-1990s. In the event, it only ever went into limited use with the Russian armed forces and the police, and unlike Kalashnikov, Nikonov did not become a household name.

The AK-47 series led to a whole range of 7.62mm Soviet small arms including the RPD and RPK light machine guns and the SVD sniper's rifle. The Soviets followed the Americans in developing a high-velocity, small-calibre cartridge. The American Armalite M16, and its successor the M4, fires a 5.56mm round and Britain, abandoning the self-loading rifle, followed suit with the same calibre for the SA80.

In fact, the last major modification to the AK-47 was the AK-74 (followed by the AK-100 series), which fires a smaller 5.45mm x 39mm round, which is the Russian equivalent to the 5.56mm x 45mm NATO standard ammunition. The AK-74 is readily identifiable by a much larger muzzle brake and a horizontal groove in the fixed wooden laminate stock (the side-folding skeleton stock is also very distinctive).

By the late 1970s the AK-74 had succeeded the AK-47 in general service with the Soviet Army and saw extensive combat in Afghanistan, though it never caught on with the Mujahideen or the Taliban due to ammunition compatibility problems with the earlier Kalashnikovs. Unlicensed copies of the AK-74 were produced by Bulgaria, China, former East Germany and Romania.

Other developments in the Kalashnikov family saw variants produced for Soviet paratroops and tank crews. The AKD assault rifle is a special folding-stock version of the AKS-74, which can be fitted with the Soviet BG-15 40mm grenade launcher. The AKS has a solid stock and can be fitted with a silencer. Soviet armour crews were issued with the compact AKR submachine gun. The RPK-74 squad automatic weapon was succeeded by the RPKS and the heavier PKM, all of which have seen service in Afghanistan's wars.

In 1991 the Izhmash factory in Izhevsk began full-scale production of the AK-74M (M = Modernizirovanniy, or Modernised), which was accepted as the standard service rifle of the Russian Federation following the collapse of the Soviet Union. This weapon features a black synthetic side-folding stock, pistol and fore grip.

The subsequent AK-100 (101–105) series based on the AK-74, which includes weapons with a variety of different calibres, were intended for the export market with the intention of appealing to AK-47 and AKM users. They look similar to the AK-74 and feature a solid synthetic side-folding stock. These are nowhere near as common as their predecessors. Libyan opposition fighters were seen sporting AK-74M/AK-103 during 2011, which presumably came from captured Libyan Army stocks. The previous year the AK-200 was unveiled again, largely as an export product.

Prior to Moscow's decision to discontinue Kalashnikov production, the weapon's world dominance had already taken a tumble in Iraq in 2007 when it was replaced by the M16. It also slipped in Libya in early 2011. Prior to the fall of Colonel Muammar Gaddafi, Russia was contracted to provide billions of dollars' worth of equipment as part of the Libyan Armed Forces modernisation programme. This included a $500 million contract to build a factory in Libya to produce the AK-103, but this was disrupted by the revolution and the death of Gaddafi. Only Venezuela and Kazakhstan and possibly India now have licences to produce the Kalashnikov legally.

The Kalashnikov has also disappeared from service in China. The Type 81 replaced the Chinese Army's Type 56 in the 1980s and this in turn was superseded by the QBZ-95, a completely new design bullpup rifle, in the mid-1990s. The Type 81 (which draws on the AK-47, the SKS and the Dragunov) has been exported but is not in use in anything like the numbers of the Type 56. Similarly the QBZ-95 is only in use with about half a dozen countries. In contrast, it is estimated that up to fifteen million Type 56s were produced, accounting for around one-fifth of all AK-47s in circulation.

Similarly the former Yugoslavia's M70 has since been replaced by the Serbian M21, which came into service in 2004 and has seen limited combat in Iraq. Compared to the M70, exports of the M21 have been very limited. Much to Russia's displeasure, Bulgaria's Kazanlak persists in marketing the AR-M range of assault rifles based on the Kalashnikov for the international export market. Similarly Iran markets its KL for export.

Although Egypt's army is now largely armed with American equipment, the Maadi Engineering Industries Company still offers customers the Misr assault rifle and semi-automatic rifle AKM clones. The Egyptian Army employs the Misr alongside the American M16 and M4. For a while in 2011 it looked as if the Egyptian armed forces might deploy their Misr to shoot large numbers of Egyptian civilians clamouring for democracy. In the end the Egyptian military chose to force President Hosni Mubarak to step down.

What does the future hold for this sixty-year-old weapon? While it is, or has been, manufactured in at least fourteen countries (including Albania, Bulgaria, China, Egypt, Germany, Hungary, India, Iraq, North Korea, Poland, Romania, Russia, Serbia and Venezuela), AK-47 derivatives are listed in over eighty countries' state arsenals. It has been assessed that the Kalashnikov 7.62mm x 39mm assault rifle will remain the dominant small arm in many parts of the world for several decades to come.

Russia reportedly has ten million Kalashnikovs in storage for an army with a strength of one million men. The worry is that many of these will end up on the international arms market to fuel yet more regional conflicts. In the meantime Izhmash pledged to provide the Russian government with a new assault rifle, but it is highly unlikely it will ever be as successful as the Kalashnikov.

The hugely successful AK-47's position as the dominant weapon of the later twentieth century remains unassailable: up to a hundred million have been produced, more than any other post-Second World War small arm. One thing is for certain: in one form or another, the compact assault rifle with the distinctive banana-shaped magazine is going to be in service for some time to come. Its combat days are far from over.

A Russian naval infantryman on exercise in 2003 in Poland with an AK-74M, the successor to the earlier AK-74. Although this rifle has been the official Russian infantry weapon for two decades, the numbers in circulation are tiny compared to the AK-47 and AKM. (*US DoD*)

Another Russian soldier, this time armed with the official Kalashnikov successor, the AN-94; this one is fitted with the Kobra sight and GP-25 grenade launcher. Although it was declared the Kalashnikov's official replacement in the mid-1990s, it was only deployed in fairly limited numbers by the Russian security forces. (*Via Author*)

A grenade launcher always helps an AKM pack an additional punch. This Afghan policeman is on patrol in Sakari Bagh, Arghandab District, Kandahar, Afghanistan, in August 2010. (*ISAF*)

These Afghan border guards have just returned from patrol to Combat Outpost Herrera, Paktiya Province, in October 2009. The man on the left is holding a Hungarian-made AMD-65. (*US DoD/SSgt Andrew Smith*)

This Afghan soldier on parade at Camp Shaheen is from the 5th Battalion, 1st Brigade, 209th Corps. His personal weapon is a Chinese Type 56 that at some stage has lost its fold-out bayonet (inevitably they get misused and break off). The stock has also undergone some unusual customising, with sections having been taken out at top and bottom. (*ISAF*)

This squatting Afghan policeman is keeping watch while members of the US 2nd Stryker Cavalry Regiment question local elders in Mizan District, Zabul Province, in August 2010. The stock, pistol grip and magazine of his Chinese Type 56-1 have been decorated in coloured tape – a common practice in Afghanistan. (*US DoD/Senior Airman Nathanael Callon*)

These Afghan soldiers are serving with the 2nd Battalion, 1st Brigade ANA. The two visible firearms are AKMs; from a distance it is impossible to date them, but they could easily be fifty years old. (*US DoD/Cpl John Scott Rafoss*)

Taliban weaponry in the town of Khar Bolaq awaiting destruction. These guns were captured by elements of the US 82nd Airborne Division during Operation Crackdown in April 2003. In the foreground are RPG grenades (minus their screw-on fin stabilisers), RPG launchers, several bolt action rifles, and a pile of AKMS and AKM assault rifles. (*US DoD/Spc Jerry T. Combes*)

As part of the reintegration and reconciliation programme, insurgents hand in their trusted Kalashnikovs at Qal'eh-ye Now, Badghis Province capital, Afghanistan, in August 2010. Black turbans are normally associated with the Taliban. *(ISAF)*

The surrendered arms include AKMS and AMD assault rifles, as well as a solitary rocket-propelled grenade! *(ISAF)*

It is evident from the insurgents' expressions that they are none too happy with the process. Many of these guns are probably family heirlooms from the Soviet–Afghan War or the Afghan Civil War. (*ISAF*)

A fine portrait of an Iraqi soldier from the Iraqi 6th Army Division on a combined patrol with American troops in Baghdad in the summer of 2007. The ribbed receiver cover and receiver dimple above the magazine identifies his small arm as an AKM – this is the most common type of Kalashnikov in circulation, even though the design dates from the late 1950s and early 1960s. (*US DoD/SSgt Bronco Suzuki*)

These Iraqi policemen are patrolling to safeguard the streets of Rashid in the summer of 2007 with men from the US 2nd Infantry Division. The weapon on the left is another AKM, while the man on the right has an equally elderly Bulgarian AKS-47, recognisable by the milled receiver and the mottled plum-coloured plastic fore grip. (*US DoD/SPC Elisha Dawkins*)

An American staff sergeant from the US 4th Infantry Regiment does his utmost to drum home the importance of providing rifle cover during house-clearance training in Baghdad. The Iraqi policeman has raised and aimed his old AKM ready to engage. (*US DoD*)

More Iraqi police under the watchful eye of American instructors from the US 7th Cavalry Regiment in Taji. Although the Iraqi Army's Kalashnikovs have been replaced by the M16, Iraq's police remain reliant on AKM hand-me-downs. (*US DoD/Senior Airman Steve Czyz*)

In Afghanistan and Iraq the Kalashnikov is not renowned for its accuracy (in part due to poor maintenance, the general age of the weapons and the quality of the ammunition) and therefore basic marksmanship training is vital for the Afghan and Iraqi police. (*US DoD/Spc Jordan Heuttl*)

This Afghan policeman is practising aiming his AMD-65 while on the move. His unit is the 7th Battalion of the Afghan Border Police, based at Garmsir – the scene of bloody battles between the British Army and the Taliban. (*US DoD/LCpl Dwight Henderson*)

Iraqi troops of the 1st Army Division on parade in Habbaniyah with their AKMs just before America decided to replace all Iraq's Kalashnikovs with M16 assault rifles and M4 carbines. This, like the Russian decision not to order any more AK-74s for the Russian armed forces, will be seen as a turning-point in the weapon's long history. (*US DoD*)

Time for reconstruction? The Kalashnikov has been a key player in the world's regional conflicts for over sixty years. The wars fought in Afghanistan probably best exemplify its remarkable durability. Nor has any man's name ever been so inextricably linked to his creation. For Mikhail Kalashnikov it has been a mixed blessing – all that can be said is that he has not profited from the AK-47, or from the blood that it has spilt. (*ISAF*)